Lecture Notes in Control and Information Sciences 330

Editors: M. Thoma, M. Morari

Petros Koumoutsakos, Igor Mezic (Eds.)

Control of Fluid Flow

Springer

Editors

Petros Koumoutsakos

Professor of Computational Science
ETH Zürich
CH-8092 Zürich
Switzerland
E-mail: petros@ethz.ch

Igor Mezic

Department of Mechanical Engineering
University of California Santa Barbara
Santa Barbara, CA 93106
USA
E-mail: mezic@engineering.ucsb.edu

Library of Congress Control Number: 2006920195

ISSN print edition: 0170-8643
ISSN electronic edition: 1610-7411
ISBN-10 3-540-25140-5 Springer Berlin Heidelberg New York
ISBN-13 978-3-540-25140-8 Springer Berlin Heidelberg New York

Springer is a part of Springer Science+Business Media
springer.com

Printed in The Netherlands

Typesetting: by the authors and TechBooks using a Springer LaTeX macro package

Cover design: *design & production* GmbH, Heidelberg

Printed on acid-free paper SPIN: 10765791 89/TechBooks 5 4 3 2 1 0

Preface

Flow Control and Optimization is increasingly incorporated in the design of engineering devices ranging from aircraft and space vehicles to microfluidic devices. Research in this field aims to integrate knowledge in flow physics with theoretical advances and novel experimental capabilities and engineering devices capable of materializing advanced design concepts. In turn controlled flow phenomena exhibit a new arrays of physics and open frontiers for further interdisciplinary research in fluid mechanics and all the engineering domains that it affects.

This collection of papers in flow control, showcases representatives lines of work in the area of flow control demonstrating the interdisciplinary character and high scientific merit of this field. The authors of these articles are among the pioneering researchers in this field and continue to explore the frontiers in flow control and optimization. We hope that this volume, while largely representing work by the authors prior to 2004, contributes in providing an authoritative report on this rapidly changing field.

Zürich, Switzerland
Santa Barbara, CA, USA
February 2006

Petros Koumoutsakos
Igor Mezic

Acknowledgments

We wish to thank all authors for their active participation and their cooperation throughout the completion of this volume. Thomas Ditzinger from Springer helped the efficient completion of this project. Professors Fernholz was instrumental in initiating this project and in providing valuable guidance throughout the assembly of the volume. He is one of the foremost pioneers in flow physics and in flow control and we wish to dedicate to him this volume to him on the occasion of his retirement.

Contents

List of Symbols

A	excitation amplitude ($= v_{rms}/U_N$)
b	slot width
c	chord length
c_1, c_2	constants
c_D	drag coefficient
c_f	mean skin-friction coefficient ($= 2\overline{\tau_w}/(\rho U_\infty^2)$)
c_f'	fluctuating skin-friction coefficient ($= 2\sqrt{\overline{\tau_w'^2}}/(\rho U_\infty^2)$)
c_L	lift coefficient
c_{L_0}	lift coefficient without forcing
c_p	static pressure coefficient ($= 2(p(x) - p_{ref})/(\rho U_\infty^2)$)
$\tilde{c}_p$	sound-pressure coefficient ($= p_{rms}/(\rho U_\infty^2)$)
c_μ	oscillatory momentum coefficient ($= \rho b v_{rms}^2/(\rho c U_\infty^2/2)$)
c_{μ_h}	c_μ based on obstacle or step height ($= \rho b v_{rms}^2/(\rho h U_\infty^2/2)$)
E_{coh}	coherent kinetic energy
f	frequency, frequency of the forcing
f_0	excitation frequency f_{exc} or corresponding subharmonic f_{sub}
f_{exc}	excitation frequency
f_{fun}	fundamental frequency
f_{sub}	first sub-harmonic frequency ($= f_{opt}/2$)
F, F^*	functions
h	height of an obstacle, step height
h_f	height of surface fence
H	tunnel height above the test section wall
H_{12}	shape parameter ($= \delta_1/\delta_2$)
l_S	length of the separation region
l_{SF}	distance between the fence and the actuator
L	characteristic length
L_f	distance between actuator and trailing edge of an airfoil
L_p	sound-pressure level

p	static pressure
p_{rms}	rms-sound pressure in the pressure chamber
Re	Reynolds number ($= U_\infty L/\nu$)
S_{uu}	power spectral density of the u' fluctuation
St	Strouhal number ($= fL/U_\infty$)
t	time
Tu_δ	free-stream turbulence level ($= \sqrt{\overline{u'^2}}/U_\infty$)
u	velocity in free-stream direction
u'	fluctuating velocity in free-stream direction
$\langle u \rangle$	phase-averaged velocity in free-stream direction
u_c	convection velocity
u_{rms}	rms-velocity in free-stream direction
U	mean velocity in free-stream direction
$U_{\max}$	maximum velocity of the profile
U_N	velocity in the free-stream normal to the fence
U_∞	free-stream velocity
v	velocity component normal to the wall
v_{rms}	rms-velocity at the slot exit
w	length in spanwise direction
x	coordinate in streamwise direction
x_A	x-coordinate of the actuator position
x_D	x-coordinate of the separation line
x_f	x-coordinate of the fence
x_R	x-coordinate of the reattachment line
x_{R_0}	x-coordinate of the reattachment line without forcing
x_{slot_1}	streamwise position of the upstream slot
x_{slot_2}	streamwise position of the downstream slot
y	coordinate normal to the wall
z	spanwise coordinate
α	angle of incidence
α_{lim}	angle of incidence when separation line coincides with actuator position
α^*	sweep angle
δ	boundary layer thickness
δ_0	boundary layer thickness at the edge of a step
δ_1	displacement thickness
δ_2	momentum-loss thickness
Δc_L	change of the lift coefficient c_L
Δf	frequency bandwidth
ΔH^*	height of a spoiler above the wall
Δx_S	length of the closed reverse-flow region ($= x_R - x_D$)
$\Delta\varphi$	phase difference
ν	kinematic viscosity
ρ	density

τ_w	wall shear-stress
φ	phase angle
χ_w	reverse-flow factor at the wall
$\langle\omega_z\rangle$	phase-averaged spanwise vorticity component

Subscripts

b	quantity based on slot width
c	quantity based on chord length
coh	quantity referring to coherent structures
h	quantity based on obstacle or step height
l_S	quantity based on length of separation region
max	maximum value of a quantity
opt	optimal value of a quantity
ref	reference value of a quantity
δ_2	quantity based on boundary layer momentum thickness

Control of Weak and Strong Reverse-Flow Regions in Incompressible Turbulent Boundary Layers

Hans-Hermann Fernholz and Frank Urzynicok

Hermann-Föttinger-Institut, Technische Universität Berlin

Abstract. A specific aspect of flow control is the manipulation of flows with open separation or with closed reverse-flow regions. These reverse-flow regions can be classified as weak or strong [19] depending mainly on the magnitude of the mean skin friction. This article does not attempt to discuss the more important investigations published in the literature but is confined to work performed by the boundary-layer group at the Hermann-Föttinger-Institut of TU Berlin over the last ten years, work which was often published only in conference proceedings or students' dissertations. The test cases chosen will be discussed in the context of whether manipulation is applied to weak or strong reverse-flow regions. In the former case separation can be eliminated completely (pressure driven separation regions) whereas in the latter case (fixed separation) the separation region can only be reduced. Weak reverse-flow regions occur mainly in external flows (e.g. suction side of an airfoil) or in internal flows (e.g. diffuser flows), strong reverse-flow regions upstream and downstream of an obstacle or in connection with a trapped vortex.

Both passive and active devices were used for the manipulation of the various separated flows, such as vortex generators and turbulence generating grids as well as mechanical spoilers and zero net-mass flux actuators consisting of a loudspeaker-pressure chamber-slot system. Measuring techniques used were hot-wire and pulsed-wire anemometry, LDA and PIV, with special emphasis on measurements of the mean and fluctuating skin friction.

1 Introduction

When Fiedler & Fernholz [22] wrote their review on "management and control of turbulent shear flows" their intention had been to draw the attention of a wider group of engineers to turbulent flow control in order to speed up the transfer of knowledge from aerodynamics to applications in other fields of engineering. This is still a worthwhile goal.

Since 1990 several good reviews have been published, e.g. Gad-el-Hak & Bushnell [24], Wygnanski [68] and Greenblatt & Wygnanski [26], covering a wider field than in the present paper. The authors confine themselves therefore to work performed by the boundary-layer group at the Hermann-Föttinger-Institut of TU Berlin over the last ten years which has generally been published only in conference proceedings or students' dissertations. The results of separation control will be shown by a demonstration of their effect on two kinds of separation regions which are characterized by weak or strong reverse flow [19].

Weak reverse flow (WRF) generally occurs if a boundary layer separates from a smoothly contoured wall due to a sufficiently strong adverse pressure gradient. Typical examples of WRF-regions can be found in diffusers or on the suction side of thick airfoils at high angles of incidence. The flow may form an open separation region or reattach at the wall, generating a closed reverse-flow region or bubble which is often steady in the mean. It is characterized, for example, by streamwise distributions of slowly rising static pressure and almost constant but small values of negative mean skin friction. The separation process depends on the upstream history of the flow and the boundary conditions in the streamwise direction. Both the separation and the reattachment line, if present, are free, i.e. their locations x_D and x_R vary instantaneously but are steady in the mean. The distance $\Delta x_S = x_R - x_D$ is one of the characteristic lengths of the problem. Manipulation may move x_D downstream or x_R upstream or eliminate both, leading to fully attached mean flow.

Strong reverse flow (SRF) occurs if a boundary layer separates from a sharp edge - be it a backward facing step or a fence. The separated shear layer reattaches, if conditions are favorable, covering a cavity-like flow region. It is characterized by large spatial changes in static pressure and mean skin friction and by large values of the fluctuating skin friction distribution. The curved separated shear layer, often convex, is practically independent of its upstream history but depends on the flow geometry and the boundary conditions. The location of the separation line, x_D, is fixed and the only variable is the location of the reattachment line, x_R, which can, however, be steady in the mean. Here the goal of the manipulation is to move x_R upstream and to shorten the reverse-flow region but it is not possible to eliminate the reverse flow completely.

The goal of manipulation or, even better, closed-loop control of reverse-flow regions is to reduce their extent or to eliminate the reverse flow. This can increase the lift and reduce the drag of an airfoil or reduce the noise and vibrations caused by separation. Overall we aim to improve the performance of flow configurations in the fields of aerodynamics, fluid machines or chemical process engineering. The relationship between separated flow regions and flow unsteadiness, respective noise, is well known but often overlooked.

As early as 1904 Prandtl demonstrated the control of the separation of the boundary layer in a diffuser by steady suction and steady blowing has been applied for a long time to prevent separation on airfoils or for cooling purposes on turbine blades. Today, unsteady blowing is a more cost-effective technology and, what is just as important, feasible. It was first applied by Seifert et al. [49] and by Seifert & Wygnanski [50] who replaced steady tangential blowing on an airfoil flap by unsteady blowing. A second technique, combining unsteady suction and blowing (e.g. [66]), acts as a zero net-mass flux (ZNMF) actuator using the fluid of the boundary layer itself without the need for a separate source of fluid. This unsteady manipulation technique either generates or manipulates coherent structures. It increases entrainment using the Kelvin-Helmholtz instability mechanism in the outer part of a separated flow, i.e. in the separated shear layer, or generates vortices in the near-wall region of a separating boundary layer and thus adds kinetic energy. Depending on the type of actuator the generated vortices are oriented in the spanwise or in the longitudinal direction.

Longitudinal vortices are generated, for example, by solid vortex generators (passive) or by vortex generator jets (VGJ). The classical vortex generator (e.g. [60], and [48]) enhances mixing and thus increases streamwise momentum near the wall by shedding longitudinal vortices but increases the drag of the whole configuration under conditions when separation is absent. This latter effect can be reduced if micro-vortex generators are used [40]. VGJ or pneumatic turbulators [28] are active actuators where the strength of the steady jet can be controlled. They have the advantage that the drag associated with fixed vortex generators is avoided with the jet flow turned off [11]. In this latter investigation "weak" (see also [36]) longitudinal vortices are generated by pitched and skewed jets. Urzynicok & Fernholz [63] discuss also a third device, an unsteady vortex generator, to manipulate separation regions by longitudinal vortices.

The second group of coherent structures which can be used for manipulating separation are spanwise vortices. They exist naturally in free and separated shear layers or can be generated by unsteady blowing and suction through a spanwise slot embedded in the wall. The slot, a pressure chamber and a piston (e.g. a loudspeaker membrane or a piezoelectrically driven metal diaphragm) form a zero net-mass flux (ZNMF) actuator. The frequency and amplitude of the perturbation can easily be adjusted to the conditions of the main flow so that the free shear layer or boundary layer responds to its excitation. Such actuators - sometimes called synthetic jet actuators ([59, 6], for example) - are small, light, energy efficient and independent of propulsive systems [68]. They were used by Ahuja & Burrin [1] demonstrating that, at least at low Reynolds numbers, separation on an airfoil can be controlled by - what they claimed to be – internal acoustic excitation (see also [31], and [29]). Williams et al. [66] emphasized that the velocity fluctuations at the slot exit due to the pumping effect of the loudspeaker can be several orders of magnitude larger than those generated by the acoustic wave. Today with understanding of the spanwise vorticity generation (e.g. [14], and [17]) the term "fluidic" excitation or forcing seems to be more appropriate (see also [3]).

The amplification of the imposed perturbation takes place primarily in the downstream shear layer rather than in the upstream boundary layer which, as shown by stability analysis [69], is insensitive to the detail of the imposed perturbations.

At higher Reynolds numbers transition effects in the separated shear layer can no longer serve as an explanation for a delay of separation. Béraud [4] and Erk [17] visualized the interaction between the unsteady flow from the slot and the cross-flow boundary layer in a water tunnel and found that the spanwise vortices generated at the sharp edges of the slot orifice of the ZNMF-actuator were responsible for re-energizing the near-wall region of the boundary layer. This is similar to the vortex generation mechanism of vibrating flaperons ([39], and [42]) or oscillating spoilers ([7, 21, 67], and [53], for example). Within the range of frequencies investigated, i.e. below 50 Hz, the spoiler type actuators are equivalent in their effects to the ZNMF-actuators ([53], and [61]).

ZNMF-actuators can also be driven solely by energy from the boundary layer itself if the slot-pressure-chamber configuration is designed according to the principles of a Helmholtz resonator ([17, 23], and [63]).

Depending on the spanwise extent of the slot [63] the ZNMF-actuator generates spanwise vortices or, if short and driven in antiphase, produces longitudinal vortices due to the shear layer interaction between the slot sections.

For the manipulation of weak and strong reverse-flow regions we shall use here only ZNMF-actuators. They increase momentum in the near-wall region of a boundary layer with incipient separation or enlarge entrainment of the outer shear layer and thus move reattachment upstream. The actuators are embedded in a cavity of the wall with the slot exit normal to the wall and they use loudspeaker membranes as pumping devices.

Before dealing with some specific cases, five items should be heeded which are a prerequisite for the manipulation of wall-bounded shear flows:

(i) What is the goal of the manipulation of the separation region?
  (a) Maximum effectiveness of the manipulation, e.g. energy input lower than energy gain,
  (b) minimum length of the separation region,
  (c) maximum lift to drag ratio,
  (d) maintaining lift beyond the natural stall angle or
  (e) reduction of noise or vibrations.
(ii) Which flow region is most suitable for the manipulation?
(iii) Manipulation by active or passive actuators?
(iv) What are the optimum properties of the actuator signal (wave form, frequency, amplitude and phase angle) for a specific shear layer ?
(v) How is the success of the manipulation controlled ?

In the cases discussed below, manipulation had mainly the goal to reduce the length of the separation region as much as possible, for example by free-stream turbulence (passive), or to increase lift in the post-stall regime on an airfoil by means of a ZNMF-actuator (active) regardless of the energy input. In the latter case the flow region most sensitive to the actuator signal must be determined in order to achieve the highest amplification of the manipulation. For a WRF-region this location lies slightly upstream of the mean separation line in a region where there is still instantaneous reverse flow. For a SRF-region the actuator was located at or close to the fixed separation line which in general is identical with the origin of the separated shear layer. If this was set, only the optimum properties of the actuator signal had to be determined. Finally, the success of the manipulation was controlled by measurements of either the skin friction, in order to determine separation and reattachment lines, or of lift and drag.

For the solution of a control problem it is advantageous to determine the variables by dimensional analysis. For a WRF region (e.g. on a given airfoil) the dependent variables are primarily functions of (e.g. [17]):

$$F\left(St_c, Re_c, c_\mu, \alpha, \frac{b}{c}, \frac{x_A}{c}, \frac{w}{c}, \text{wave form of the excitation}\right).$$

Here the following definitions are used: The Strouhal number $St_c = fc/U_\infty$ (with a different set of fundamental parameters it could be defined with the slot width b as $St_b = fb/U_\infty$), the Reynolds number $Re_c = cU_\infty/\nu$ where c is the chord length and U_∞ the free-stream velocity, the momentum coefficient $c_\mu = \rho b v_{rms}^2/(\rho c U_\infty^2/2)$ which is sometimes replaced by v_{rms}/U_∞ (with v_{rms} as the rms velocity at the slot exit) or a sound-pressure coefficient $\tilde{c}_p = p_{rms}/\rho U_\infty^2$ with p_{rms} as the rms-sound pressure in the pressure chamber, the angle of incidence α, the position of the actuator x_A/c and the aspect ratio of the airfoil w/c. For a boundary layer along a flat plate Re_c would be replaced by Re_{δ_2} with δ_2 as the momentum-loss thickness. Wygnanski [68] defines St by using as a length scale the distance between the actuator and the trailing edge of the airfoil, L_f. A note of caution regarding the sound-pressure coefficient $\tilde{c}_p$ is necessary here. Since there is often a rather complicated transfer function between the sound-pressure level in the pressure chamber and the velocity perturbation at the slot exit, a parameter using v_{rms} is more suitable (as suggested by [52], and [66]), especially since v_{rms} can now be measured by means of LDA, for example.

For a SRF-region, e.g. downstream of a fence, dimensional analysis (e.g. [53]) shows primarily a functional relationship:

$$F^*\left(St_h, Re_h, c_{\mu_h}, \frac{x_f}{h}, \frac{b}{h}, \frac{H}{h}, \frac{w}{h}, \frac{x_A}{h}, \alpha^*, \text{wave form of the excitation}\right).$$

Deviating from Siller, we have introduced $c_{\mu_h} = \rho b v_{rms}^2/(\rho h U_\infty^2/2)$. Again one could also have chosen the amplitude parameter v_{rms}/U_∞ or the parameter $b v_{rms}^2/(h U_\infty^2)$ [47], instead. St_h is defined as fh/U_∞ but could also be formed with the slot width depending on considerations of the far or the near field. The Reynolds-number is $Re_h = hU_\infty/\nu$ and x_f the position of the fence which could be replaced by the boundary layer thickness δ at the fence without the fence being present, H the height of the test section, l_{SF} the distance between the fence and the actuator if it is upstream of and not at the tip of the fence and α^* the sweep angle of the fence.

This latter set of parameters is generally used in the literature but has the disadvantage that at least three parameters change with a variation of U_∞ and an optimum value of the derived dependent variable Δx_S cannot be stated a priori. This can be avoided [53] if h, ν and ρ are used as the natural dimension base and this then gives for the non-dimensional frequency fh^2/ν and $\tilde{c}_p$ as $p_{rms}h^2/(\rho\nu^2)$.

The relationship F^* applies also to the flow over a backward facing step if x_A is put to zero and x_f/h replaced by δ_0/h where δ_0 is a characteristic boundary layer thickness at the edge of the step. One must also account for the initial conditions of the oncoming flow, including the state of the boundary layer.

2 Weak Reverse-Flow Regions

Weak reverse-flow regions, as defined above, may be found in pressure-induced separation regions with and without reattachment. Examples are the separation bubble downstream of the sharp-edged blunt face of a circular cylinder [52] or a blunt plate [8], axisymmetric separation bubbles in tailored pressure distributions ([2, 12], and [38], for example) or separation regions on the suction side of airfoils (e.g. [17, 21, 55], and [62]). This latter configuration has probably attracted the largest number of investigations (see [22], and [26]). Furthermore there are reports on separation control in diffusers (e.g. [68], and [10]). Most of these flows have in common that the magnitude of the skin friction in the open or closed reverse-flow region is small so that it should be possible by adding enough momentum to the near-wall region to push separation downstream or to force the boundary layer to reattach if the separation region is open (e.g. [21]). If a "trapped vortex" is generated by forcing reattachment, this vortex may be so strong that the reverse flow has skin friction values large enough to change the flow from weak to strong reverse flow (see Sect. 3.2). Forced reattachment is of special importance in cases where post-stall occurs, i.e. where the boundary layer separates close to the leading edge, with the consequence that the downstream forward flow is far from the suction surface of the wing. For airfoils with a suction peak in the first quarter of the chord it may suffice to push separation so far downstream that the suction peak, reduced by leading-edge stall, is recovered and lift is regained (e.g. [17], and [62]). If separation occurs towards the trailing edge it is also possible to control the separation region (e.g. [30], and [57]) but the increase in lift is usually smaller.

Internal flow with an adverse pressure gradient can be manipulated in order to reduce the length of or even eliminate a closed reverse-flow region with the aim of increasing pressure recovery or of reducing flow unsteadiness and with it noise ([38], and [18]).

A note of caution should be expressed at the beginning of 2: Flows over wings with large separation regions show interesting physical phenomena if manipulated, depending on the type of stall, i.e. whether it is leading-edge (post-stall) or trailing-edge stall. The results are, however, often specific of an airfoil type and difficult to generalize. This will be demonstrated in the following case studies.

2.1 Manipulation of Separation on an Airfoil Under Post-Stall Conditions at High and Low Frequency Excitation

In investigations by Erk [17] and Urzynicok [61] the flow around a wing section separating at the leading edge at high angle of attack, thus forming a weak reverse-flow region, was manipulated using high- and low-frequency forcing. In both regimes ZNMF-actuators were utilized. For comparison, the measurements in the low-frequency range were complemented by applying mechanical actuation.

The experiments were carried out on a Wortmann FX 61-184 laminar glider profile with chord $c = 450\,\text{mm}$ and aspect ratio $2:1$. The angle of incidence α was $26°$ and the Reynolds number based on the chord length ranged from $Re_c = 2.5 \times 10^5$ to 1.0×10^6.

2.1.1 Separation Control Using Fluidic Forcing at High Frequencies

For fluidic forcing the ZNMF-actuator consisted of a spanwise slot ($b = 1.6\,\text{mm}$) and a spanwise circular pressure chamber with a loudspeaker attached at each end. The slot was located just downstream of the separation line at $x/c = 0.02$ (Fig. 1(a)). Forcing was performed in a frequency range between

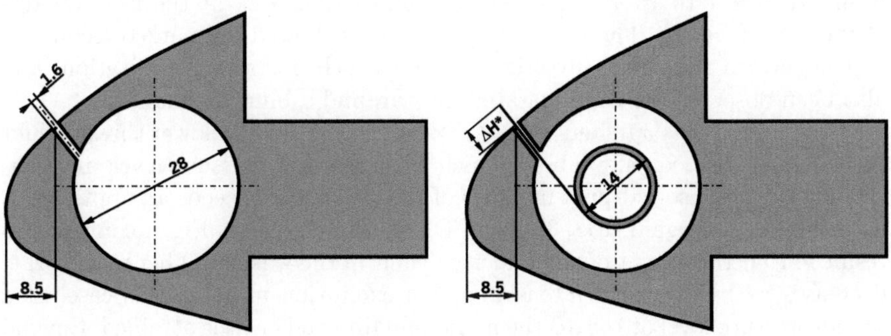

(a) ZNMF-actuator with loudspeakers at the ends. Note the slot with the cavity underneath serving as sound wave duct. From Erk [17].

(b) The mechanical actuator consisting of a spoiler with extension height ΔH^* and a driving shaft.

Fig. 1. Cross sections of the fluidic and of the mechanical actuator implemented into the leading edge section of the airfoil

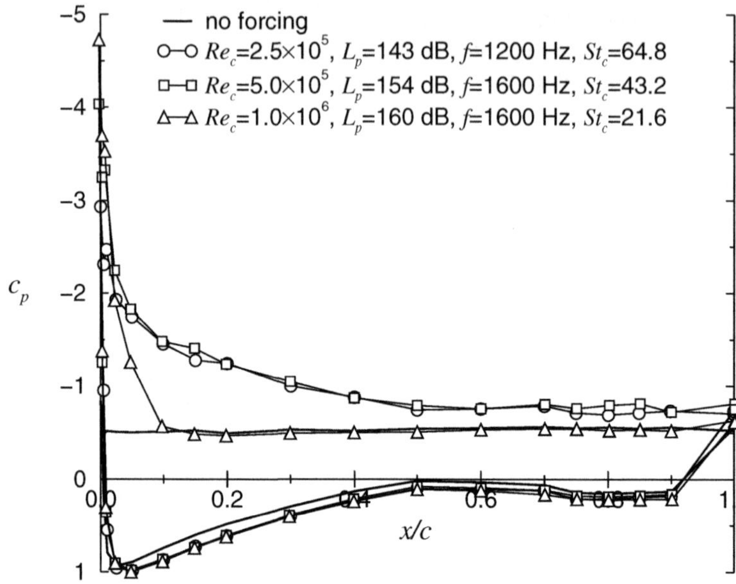

Fig. 2. Pressure distributions of the flow with and without manipulation ($\alpha = 26°$). From Erk [17]

800 and 2000 Hz at sound-pressure levels up to $L_p = 160\,\text{dB}$ (*re* 2×10^{-5} Pa). The excitation strength was monitored with a microphone at the centerline of the wing inside the cavity.

Oil-film visualizations of the manipulated flow on the suction side showed that for $\alpha = 26°$ and $Re_c = 7.5 \times 10^5$ the separation line moved downstream from $x/c \approx 2\,\%$ to $16\,\%$. This restored the suction peak at the leading edge of the wing (Fig. 2). The reattachment process showed no marked frequency dependence in the range investigated here which, for strong excitation, had also been observed for the separated flow around a blunt cylinder [51].

At low Reynolds numbers ($Re_c \leq 5{\times}10^5$), there exists a lower threshold for each excitation frequency above which higher sound-pressure levels increase the lift. Once the sound-pressure level of the excitation exceeds an upper limit, the lift decreases again (Fig. 3). The LDA measurements in Fig. 4 confirm this result in that the extent of the velocity deficit in the wake, and hence the drag, decreases with increasing intensity of the excitation until, in this case, at a sound-pressure level of 145 dB the maximum lift coefficient is attained. Beyond this threshold the wake widens again and drag increases. Both effects indicate that the process has reached saturation [15].

At higher Reynolds numbers ($Re_c \geq 5{\times}10^5$) hysteresis was observed, when part of the flow stayed attached even after the excitation had been turned off (Fig. 5). Experiments have shown that this effect is not due to turbulent tripping [16]. In this case, the actuator, consisting of cavity and slot, acts as a

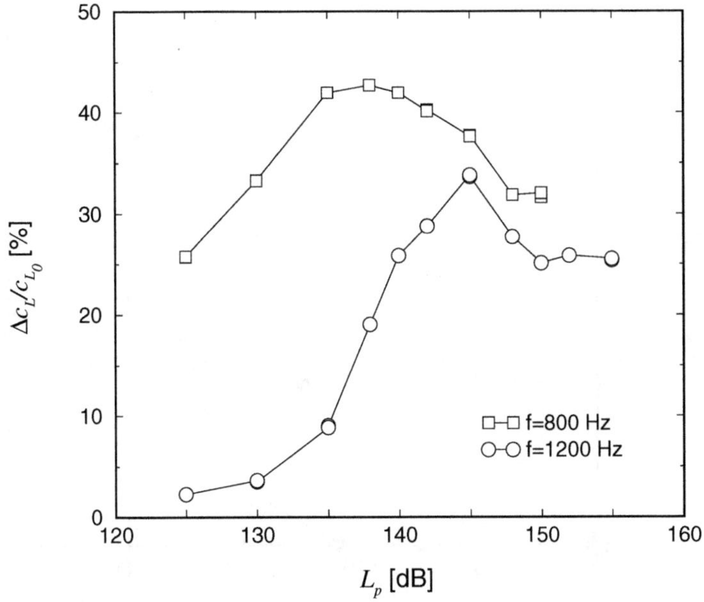

Fig. 3. Lift gain as a function of frequency and sound-pressure level ($Re_c = 2.5\times10^5$, $\alpha = 26°$). From Erk [17]

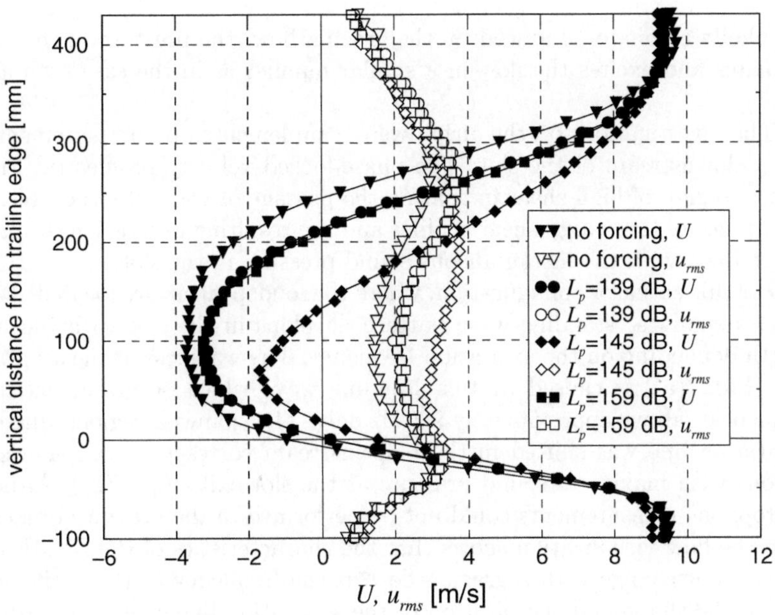

Fig. 4. LDA measurements of the velocity deficit in the wake of the airfoil with and without manipulation ($Re_c = 2.5 \times 10^5$, $f = 1200\,\text{Hz}$, $\alpha = 26°$, $x = 205\,\text{mm}$ downstream of the trailing edge). From Erk [17]

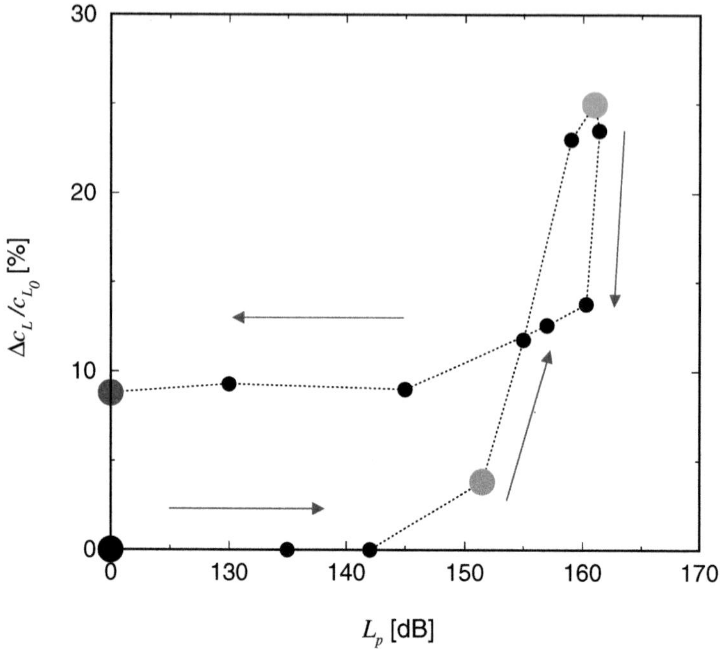

Fig. 5. Hysteresis effect ($Re_c = 1 \times 10^6$, $\alpha = 26°$). From Erk [17]

Helmholtz resonator. Induced by the grazing flow, the pressure in the cavity oscillates and excites the flow in a similar manner as in the case with active forcing.

The investigations on the airfoil were complemented by measurements directly downstream of the slot: The phase-locked velocity profiles (in Fig. 6) show a region of high shear indicating the passage of vortical structures. The occurrence of these high-shear profiles and the resulting generation of vortices are coupled with the instantaneous sound pressure in the slot.

For this actuator configuration, where the loudspeakers are located at both ends, a spanwise standing-wave pattern develops in the slot with the wavelength depending on the excitation frequency. The waviness of the separation line (Fig. 7(a)) is caused by this standing wave which occurs at excitation frequencies above approximately 900 Hz only. The spanwise regions where the separation line was shifted furthest downstream correspond to spanwise locations with maximum sound pressure at the slot exit (Fig. 7(b)). Although microphone measurements could not be performed in the presence of a crossflow, the flow visualization shows that the characteristics of the ZNMF actuator do not change with a grazing flow. If the frequency of the excitation is varied, the shape and the position of the separation line change accordingly. Adjacent peaks in the standing wave are 180° out of phase imposing a strong, oscillating, spanwise pressure gradient.

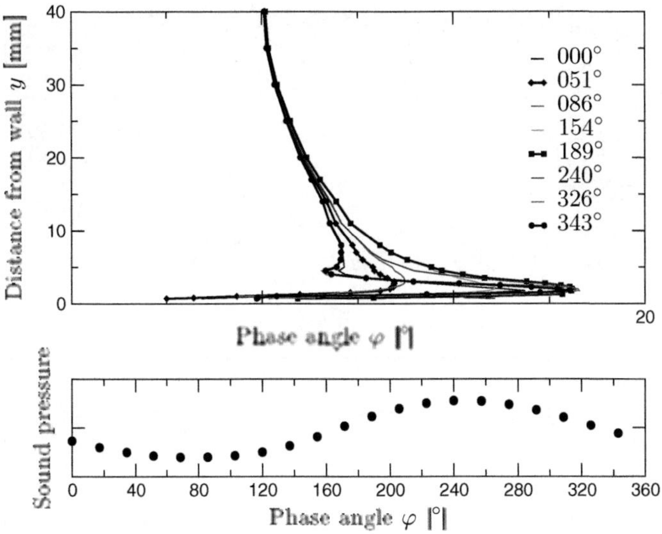

Fig. 6. Oscillating flow on the suction side of the airfoil downstream of the excitation source: Instantaneous regions of high shear in the velocity profile (top) at $y \approx 3\,\mathrm{mm}$ correlate with phase angles where the driving sound pressure inside the cavity (bottom) has a node. Phase-locked hot-wire data ($Re_c = 2.5 \times 10^5$, $L_p = 159\,\mathrm{dB}$, $f = 1200\,\mathrm{Hz}$, $x/c = 2.5\,\%$). From Erk [17]

At excitation frequencies below 900 Hz, where standing waves do not occur in the slot, the forcing signal was nearly two-dimensional which resulted in a two-dimensional separation line.

These effects imply that the flow in the geometric near field of this actuator plays an important role in determining the reattachment process. The saturation at high sound-pressure levels results from the acoustically generated unsteady jet at the slot blocking the oncoming boundary layer and, thus, promoting separation. The frequency selectivity regarding the maximum lift gain has to be attributed to the transmission properties (impedance) of the loudspeaker-cavity-slot system rather than to the receptivity of the separated shear layer.

2.1.2 Separation Control Using Fluidic Forcing at Low Frequencies

For the investigations at low excitation frequencies the same airfoil was used as described in the previous section, but the frequency of the actuator was reduced. In fact, two actuators were tested, a mechanical one (Fig. 1(b)) and a ZNMF-actuator of the same type as in Sect. 2.1.1, and it is obvious that because of the frequency limitations of the mechanical spoiler, a comparison of the two actuators could only be performed in a low-frequency range.

With fluidic excitation, forcing frequencies were in a range from 20 to 60 Hz, where the propagation of plane waves was severely attenuated. There-

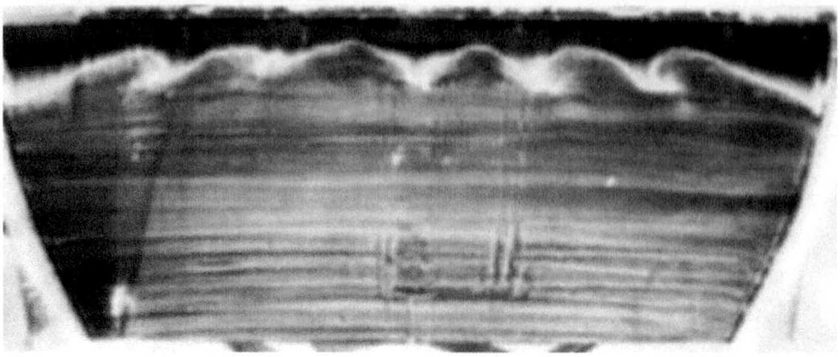

(a)

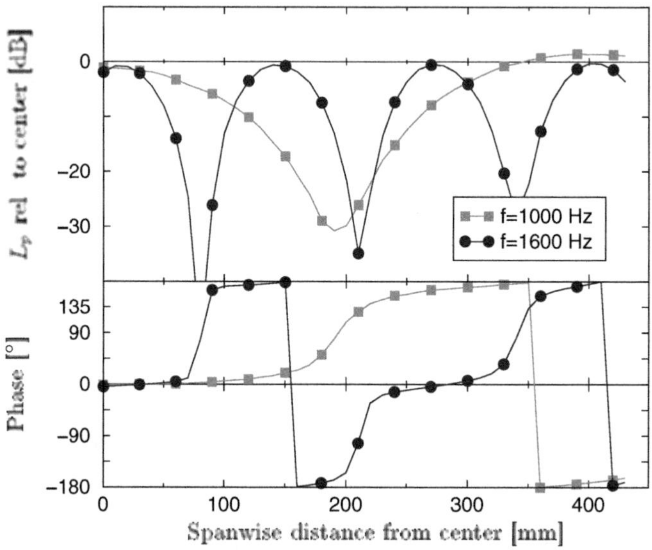

(b)

Fig. 7. (7(a)) Flow visualization showing the waviness of the separation line (flow is from top to bottom, $f = 1770\,\text{Hz}$, $L_p = 155\,\text{dB}$, $Re_c = 7.5 \times 10^5$, $\alpha = 26°$). (Fig. 7(b)) Standing-wave pattern at high forcing frequencies inside slot and cavity. From Erk [17]

fore, the excitation amplitude at the slot exit was much lower than in the high-frequency case.

Correspondingly, the effect on the separated shear-layer was less spectacular, but showed similar trends as in the higher frequency case: lift increased with a maximum gain of 13.5 % at $Re_c = 2.8 \times 10^5$, the suction peak of the pressure distribution was restored, though to a lesser extent, and velocity profiles downstream of the ZNMF-actuator were fuller, carrying more momentum in the wall proximity. Since the excitation was approximately homogeneous

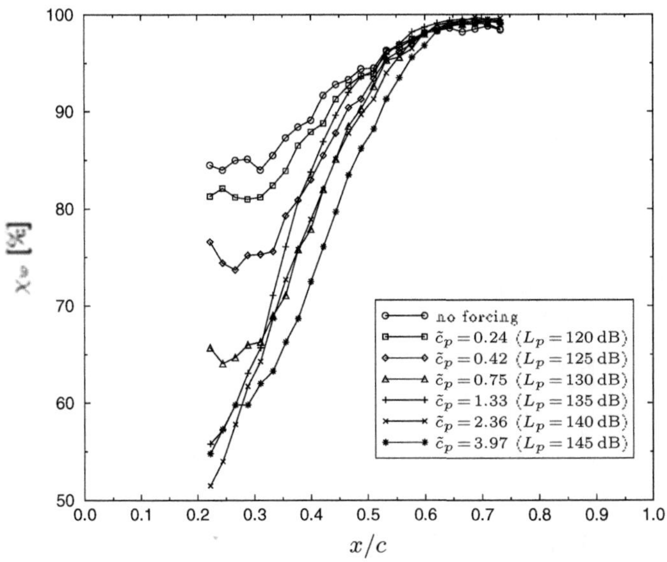

Fig. 8. Distribution of the reverse-flow factor χ_w on the suction side of the airfoil with and without low-frequency ZNMF-excitation at $St_c = 1.1$ ($f = 20\,\mathrm{Hz}$) and $Re_c = 2.5 \times 10^5$ for various amplitudes

along the span, the separation line was two-dimensional. The reverse-flow factor χ_w was reduced (Fig. 8), although χ_w stayed above 50 % at $x/c > 0.22$ indicating that on average, the flow was still separated. The forcing amplitude was given here as sound-pressure level and sound-pressure coefficient, respectively.

2.1.3 Separation Control Using Mechanical Forcing at Low Frequencies

The mechanical excitation source used in the airfoil experiment consisted of a fence-type spoiler extending along the entire span and oscillating perpendicularly to the airfoil surface. It was located at the same streamwise position as the ZNMF-actuator (Fig. 1(b)) and operated at frequencies ranging from $f = 5$ to $70\,\mathrm{Hz}$. The amplitude of the spoiler could be varied in discrete steps ΔH^* between 2, 4, and 6 mm.

Flow visualizations on the surface of the suction side of the airfoil showed that the mechanical spoiler was effective and that under optimum conditions the separation line moved downstream from $x/c \approx 0.02$ to 0.15. Flow visualizations (not shown here) in a plane normal to the suction side revealed that spanwise vorticity was generated at the edge of the oscillating spoiler having a strong influence deep into the flow field. The vortical structures shed from the oscillating spoiler, when the flow grazed past its edge, induced a downward movement which caused the flow to reattach.

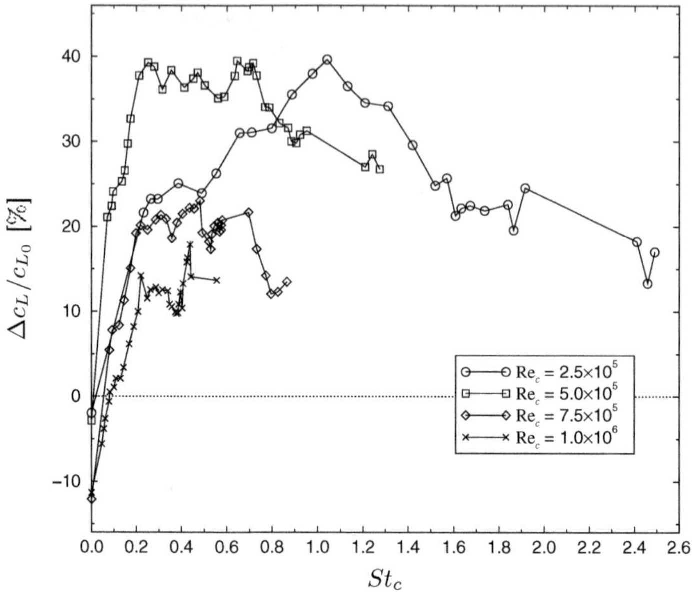

Fig. 9. Lift gain due to the mechanical oscillating spoiler as a function of Reynolds and Strouhal number (amplitude $\Delta H^* = 4\,\mathrm{mm}$). From Urzynicok & Fernholz [62]

Figure 9 shows the lift gain obtained by the mechanical spoiler as a function of Strouhal number St_c and Reynolds number Re_c. For $Re_c \geq 5 \times 10^5$ the optimum Strouhal number was about 0.2 for $\Delta H^* = 4\,\mathrm{mm}$ and for $Re_c = 2.5 \times 10^5$ St_c was about 1.05. The maximum lift gain $\Delta c_L/c_{L_0}$ was 40 %. It is interesting to note that a fixed spoiler ($St_c = 0$) had a detrimental effect. Increasing the Reynolds number beyond 5×10^5 decreased the gain in lift for this airfoil again.

Saturation, even though dependent on different parameters, was also observed in the case of mechanical forcing: Although for small amplitudes of the spoiler an increase in ΔH^* leads to an increase in lift gain, there is a threshold ($\Delta H^* = 4\,\mathrm{mm}$ in this case), above which a further increase of the amplitude did not result in an improvement. This effect might stem from the oncoming boundary layer being blocked and, thus, promoting separation again.

Measurements of the pressure distribution (Fig. 10) around the airfoil showed that the mechanical excitation partially restored the suction peak which, at high angles of attack, is mainly responsible for the lift. Naturally, the pressure distributions depend on the excitation frequency: At low Strouhal numbers relatively extended regions of low pressure occurred near the leading edge, while the suction peaks had a shorter streamwise extension but grew stronger with increasing St_c. This effect is most likely caused by the size of the vortices generated at the edge of the spoiler which scale with the excitation frequency.

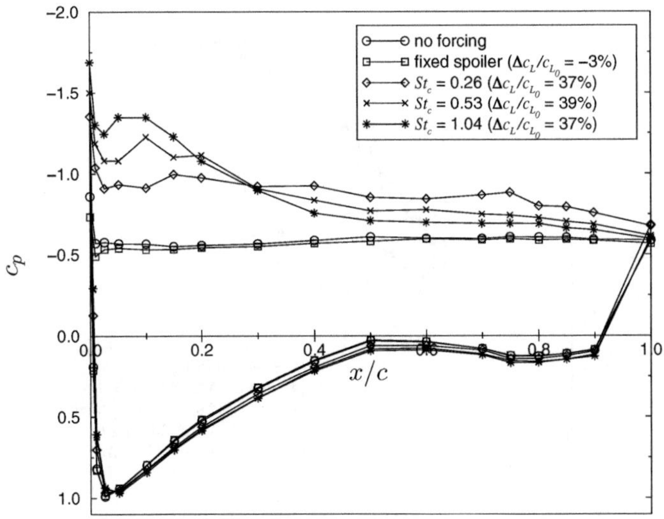

Fig. 10. Pressure distributions around the airfoil at $Re_c = 5 \times 10^5$ with and without mechanical excitation by an oscillating spoiler with an amplitude $\Delta H^* = 4\,\text{mm}$. From Urzynicok & Fernholz [62]

The reverse-flow factor, measured by a wall pulsed-wire probe, was noticeably reduced when the flow was forced (Fig. 11), but even far downstream of the separation line (up to a position of $x/c \approx 0.6$) the length of flow instantaneously attached increased significantly.

Although the results obtained with both types of actuators in the low-frequency regime vary in magnitude, the underlying effect seems to be the same. It results from the generation of spanwise vorticity. In the fluidic case, the slot acts as a "vortex pump", while in the mechanical case, intermittent vorticity is shed from the edge of the spoiler.

2.2 Manipulation of Trailing Edge Separation on an HQ-17 Laminar Airfoil by Means of Two-Dimensional ZNMF-Actuators

In contrast to case study 2.1 where we have attempted to manipulate leading-edge separation (post-stall), the goal of this investigation has been to increase lift when separation occurred in the recompression region of an airfoil. Especially in general aviation, the airfoils used are relatively thick and exhibit therefore a trailing-edge stall behavior [41]. Hence, this study aimed at manipulating the separation region before deep stall occurred at higher angles of attack. Huang et al. [30] had observed that trailing-edge separation and the wake structure responded to the excitation frequency of a ZNMF-actuator located in the recompression region if this frequency was close to the vortex shedding frequency.

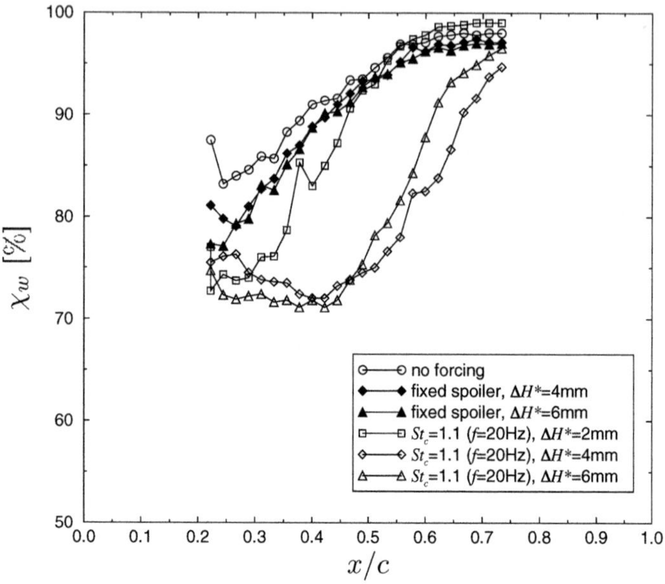

Fig. 11. Reverse-flow factor χ_w on the suction side of the airfoil with and without mechanical excitation at $St_c = 1.1$ ($f = 20\,\text{Hz}$) for various amplitudes of the spoiler ($Re_c = 2.5 \times 10^5$). From Urzynicok & Fernholz [62]

The present experiments were performed on an HQ-17 laminar airfoil (with a chord length $c = 0.65\,\text{m}$ and a span $w = 1.55\,\text{m}$) which was mounted on a six-component mechanical balance to determine the lift and drag coefficients c_L and c_D, respectively. The Reynolds number range was $5 \times 10^5 \leq Re_c \leq 1.5 \times 10^6$ and the angle of incidence $7° \leq \alpha \leq 22°$. The airfoil had two independent actuator slots of width 1.2 mm and subsequently 1.7 mm, extending over the span and fed by eight loudspeakers each. This arrangement ensured a homogeneous momentum flux along the span. For details of the experiment the reader is referred to Siller & Fernholz [55, 56, 57].

The behavior of the separation region on the wing is shown in Fig. 12 where the chord position of separation is seen to move smoothly upstream for an angle of incidence larger 7°. The respective separation angle determines the position of the actuator slot(s) which must be located upstream of the mean separation line. For the single slot configuration this position was chosen as $x_{\text{slot}_2}/c = 0.66$ at $\alpha_{lim} \approx 13°$, whereas the double-slot case had an additional upstream slot at $x_{\text{slot}_1}/c = 0.50$ (with $\alpha_{lim} \approx 16°$). Leading-edge stall would have occurred at about 22° and the separation region can be classified as weak. In contrast to a leading-edge stall scenario, the maximum lift coefficient $c_{L_{\max}}$ can be increased.

The following investigations with single and double slots were not optimization but feasibility studies meant to give indications for a parameter

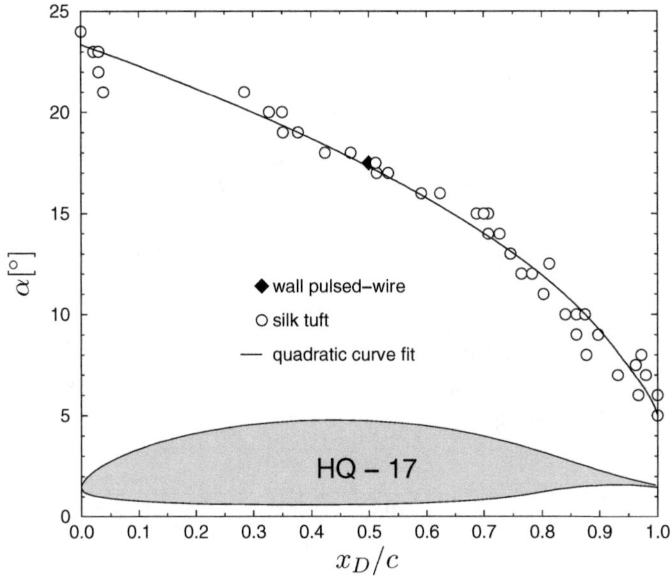

Fig. 12. Variation of the separation line as a function of the angle of incidence at $Re_c = 1 \times 10^6$. From Siller & Fernholz [55]

optimization of Strouhal number and momentum coefficient, where the number of slots and the phase difference between the actuator wave forms were additional parameters of the problem. The Strouhal number $St_{l_S} = fl_S/U_\infty$ is defined here with the length of the separation region l_S without manipulation and the momentum coefficient $c_\mu = 2bv_{rms}^2/cU_\infty^2$. With free-stream velocity U_∞, chord length c and slot width b fixed, only the frequency f and the rms-velocity fluctuation at the slot v_{rms} could be varied.

2.2.1 The Single Slot Configuration ($x_{\text{slot}_2}/c = 0.66$, $b = 1.2\,\text{mm}$)

The parameter study for the determination of the Strouhal number was performed at $\alpha = 10°$ for $1 \times 10^6 \leq Re_c \leq 1.5 \times 10^6$ and $2.3 \times 10^{-4} \leq c_\mu \leq 8.4 \times 10^{-4}$. The distribution of the lift coefficient $\Delta c_L = c_L - c_{L_0}$ (Fig. 13) shows that the optimum Strouhal number St_{l_S} is about 0.3. With $St_c = 1.43$, $c_\mu = 1.36 \times 10^{-3}$ and $Re_c = 1 \times 10^6$ c_L was measured as a function of the angle α (Fig. 14) and compared with the unforced case. The increase in c_L is about 8 % and this can also be seen in the respective pressure distributions where the manipulation does not only affect the pressure in the recompression region but also upstream (Fig. 15).

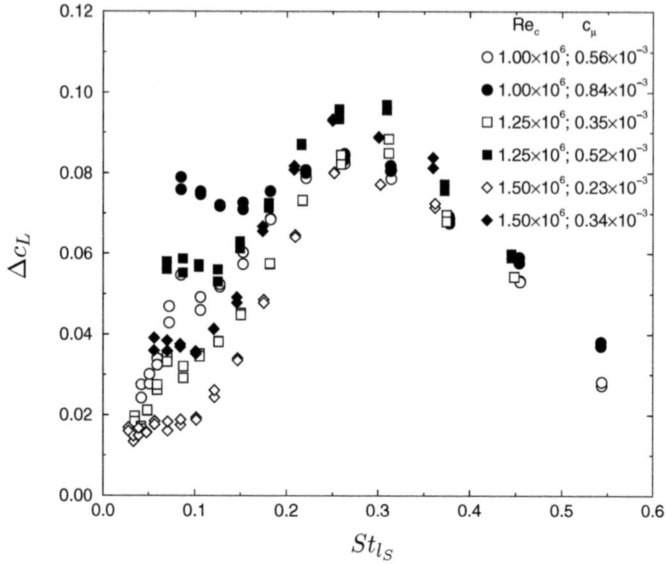

Fig. 13. Lift gain as a function of Strouhal number, Re_c and c_μ ($x_A/c = 0.66$ and $\alpha = 10°$). From Siller & Fernholz [55]

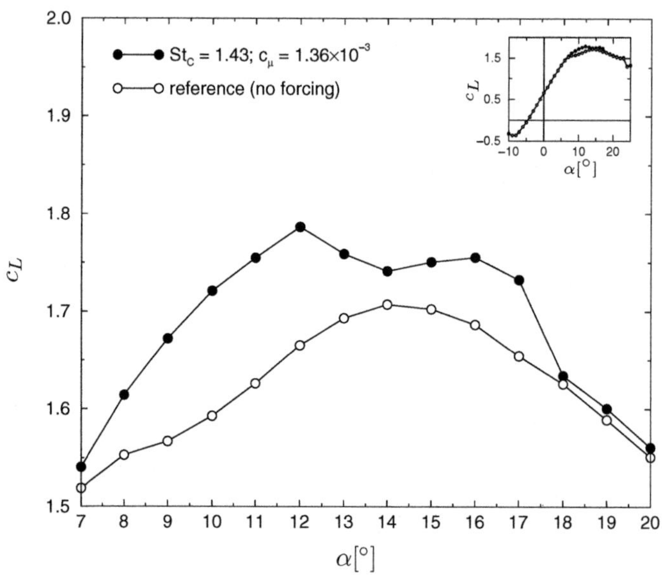

Fig. 14. Polar diagram at $Re_c = 1 \times 10^6$, $x_A/c = 0.66$. From Siller & Fernholz [55]

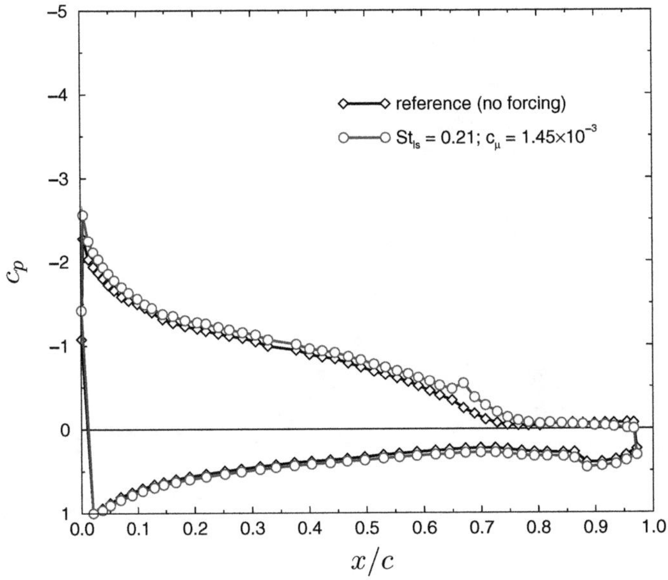

Fig. 15. Pressure distribution at $\alpha = 10°$; $Re_c = 1 \times 10^6$, $x_A/c = 0.66$

2.2.2 Double Slot Configuration ($x_{\mathrm{slot}_1}/c = 0.50$, $x_{\mathrm{slot}_2}/c = 0.66$)

Since the gain in lift for the single slot configuration was relatively small, the airfoil was provided with a second slot parallel to the first one, but further upstream ($x_{\mathrm{slot}_1}/c = 0.50$). The width of the slots was 1.7 mm. The parameter study for $Re_c = 1 \times 10^6$ and $\alpha = 12°$ showed an increase of Δc_L to about 0.12 for the single wider slot at the upstream position and a further increase, when both slots were open, to $\Delta c_L = 0.15$ (Fig. 16). Since the optimum forcing frequency is related to the shedding frequency of the airfoil, the peak values of Δc_L are achieved for forcing in a range $40\,\mathrm{Hz} \leq f \leq 50\,\mathrm{Hz}$, equivalent to $St_{l_S} = 0.3$. Figure 17 shows the influence of the manipulation on the distribution of the lift coefficient as a function of the angle of incidence α. In the experiments shown here c_μ was kept constant for both slots and the pressure distribution again reflects the higher lift (Fig. 18).

2.3 Manipulation of Closed Reverse-Flow Regions by Free-Stream Turbulence (FST)

It has been shown by Hillier & Cherry [27] that the mean flow field of a closed reverse-flow region responds strongly to free-stream turbulence intensity. They investigated a separation bubble generated at the leading edge of a blunt flat plate parallel to the flow (height $h = 38.1$ mm). Re_h was in the range $3.4 \times 10^4 \leq Re_h \leq 8.0 \times 10^4$. A free-stream turbulence level of 6.55 % generated by an upstream grid changed the non-dimensional bubble length from $\Delta x_S/h =$

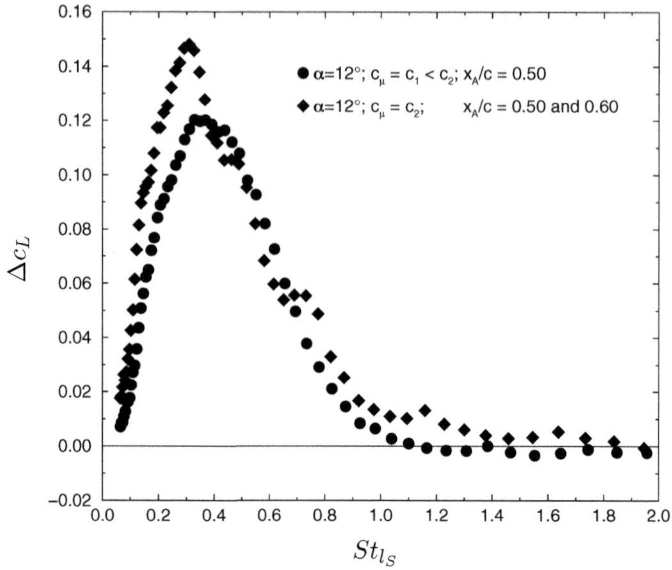

Fig. 16. Lift gain as a function of Strouhal number, number of slots and c_μ. ($\alpha = 12°$, $Re_c = 1 \times 10^6$, $x_A/c = 0.50$ and 0.60). From Zhou et al. [72]

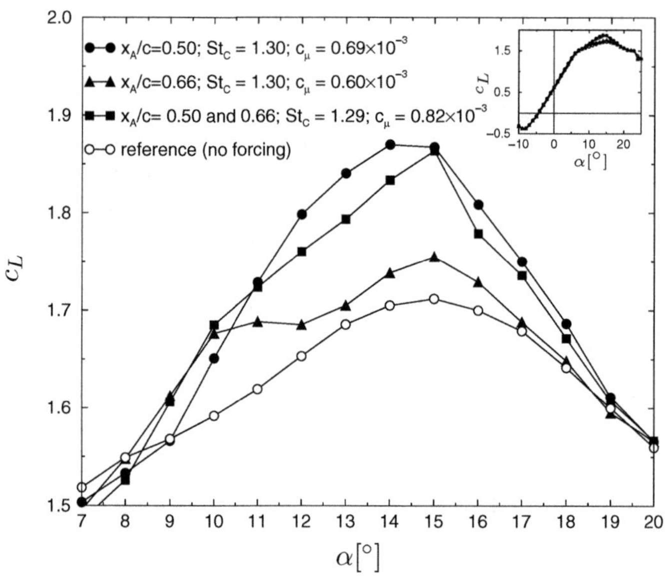

Fig. 17. Polar diagram at $Re_c = 1 \times 10^6$, $x_A/c = 0.50$ and 0.66

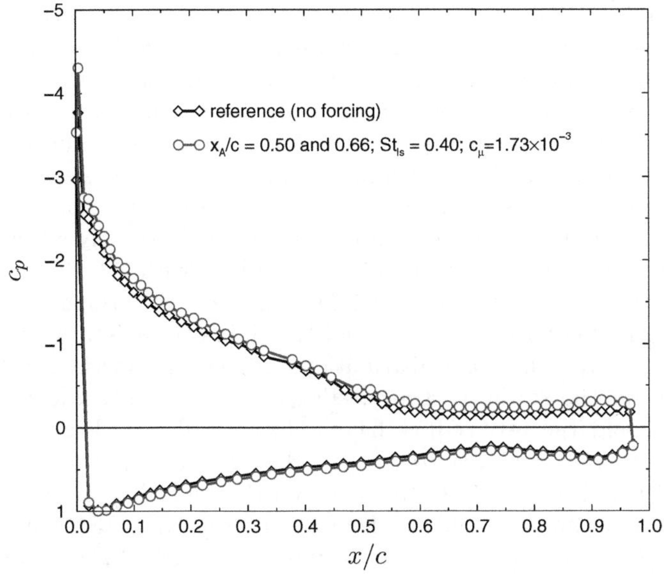

Fig. 18. Pressure distribution at $\alpha = 15°$; $Re_c = 1 \times 10^6$, $x_A/c = 0.50$ and 0.66

4.88 at $Tu_\delta = 0.2\,\%$ to 2.70, achieving a significant reduction. The negative mean pressure peak was increased by about 35 %.

A second case, where the separation line was also fixed and where free-stream turbulence changed the length of the reverse-flow region was the flow over a straight backward-facing step of height h [35]. At a step height Reynolds number $Re_h = 3.2 \times 10^4$ the non-dimensional reattachment length $\Delta x_S/h$ was reduced from 8.21 to 6.30 by increasing the free-stream turbulence intensity from 0.25 % to 7.4 % where the FST was generated by means of a coarse rectangular mesh. The mean reattachment length remained two-dimensional. The turbulence intensity near the wall at the separation line played a significant role in the development of the inner mixing layer leading to faster reattachment. So an augmentation of the entrainment rate was probably an important factor. Although skin friction was not measured in either case, the first case probably exhibits a WRF-region and the second a SRF-region. In neither case could the separation region, however, be eliminated completely. The next two cases differ in that the separation and the reattachment lines are free and the reverse-flow region is weak.

In a WRF-region the skin friction is small and negative, so that a small increase of the kinetic energy in the vicinity of the wall should suffice to change the slope of the mean velocity profile from negative to positive and thus change the position of both separation and reattachment line. If the manipulation is strong enough, the reverse-flow region can be reduced or even eliminated. This was achieved by Obi et al. [43] who used a ZNMF-actuator to suppress separation in a plane asymmetric diffuser and by Kalter & Fernholz [38] who used FST to shorten or even to eliminate an axisymmetric closed reverse-flow region in a boundary-layer flow. The turbulence intensity was varied between 0.2 % and 6 % using upstream grids while the turbulence length scale was on the order of the boundary layer thickness. Free-stream turbulence had a small effect on the boundary layer in the mild adverse-pressure gradient region but in the vicinity of separation and in the reverse-flow region mean velocity profiles, skin friction and turbulence structure were strongly affected.

Three flow cases were investigated: case LFST with 0.2 % FST (low), case MFST with 3.4 % (medium) and HFST with 5.6 % FST (high). The result of the manipulation for cases MFST and HFST in comparison with case LFST is shown in Fig. 19, where distributions in the streamwise direction x are presented for the static pressure coefficient c_p, the mean skin friction coefficient c_f, and the reverse-flow factor at the wall χ_w (denoting the probability of reverse flow near the wall). Here c_f and χ_w were measured by means of a wall pulsed-wire. LFST shows the typical c_p-distributions for a flow with a separation bubble. Mean separation ($x_D = 344\,\text{mm}$) and reattachment ($x_R = 561\,\text{mm}$) define a mean bubble length of $\Delta x_S = 217\,\text{mm}$ and are characterized as the locations where $c_f = 0$ and $\chi_w = 50\,\%$. The length of the "shoulder" in the pressure distribution corresponds roughly with the

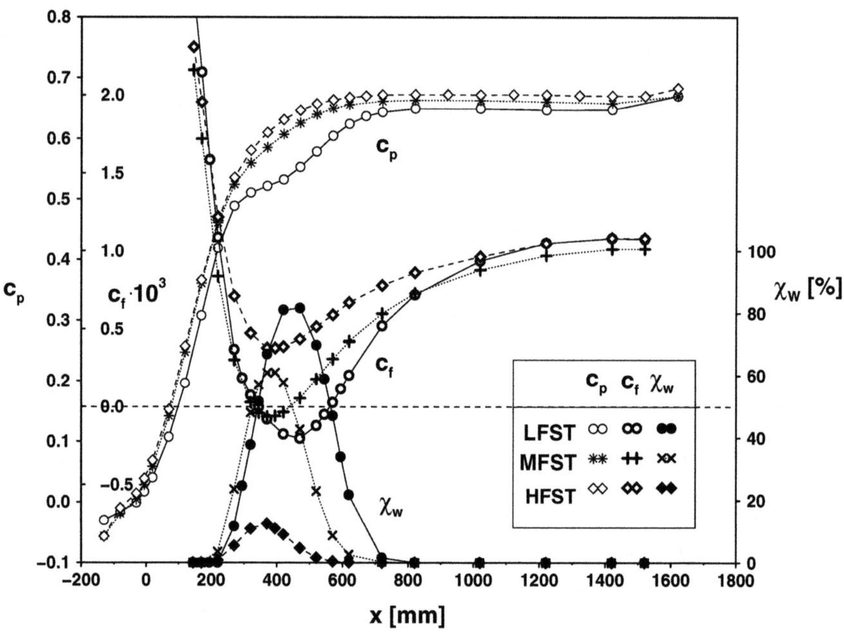

Fig. 19. Streamwise distribution of the wall parameters: static pressure coefficient c_p, skin-friction coefficient c_f, and reverse-flow parameter χ_w. From Kalter & Fernholz [38]

length of the bubble. The upstream c_p and c_f distributions are nearly identical for all three cases, showing the dominant effect of the pressure gradient until instantaneous reverse flow occurs at the wall, with χ_w increasing from zero. The c_f values for the LFST and MFST flows which will separate are virtually identical until χ_w reaches 50 %, i.e. the onset of mean flow separation. They then diverge both from each other and from the attached case HFST which lies higher throughout. In the reverse-flow region the c_f distributions are typical of WRF.

Instantaneous reverse flow was observed for LFST in a region extending from half a bubble length upstream of mean separation to one bubble length downstream of mean reattachment (see Fig. 19). This is an indicator of the "buffeting behavior" of the flow. The bubble location for LFST and MFST is well defined in the mean, however.

Since the distributions of c_p are almost identical for MFST and HFST until χ_w departs from zero, it must be the effect of the FST on the mean velocity profiles and the turbulence structure which changes the c_f and χ_w distributions when compared to those of case LFST. Although the boundary layer of case MFST still separates, its mean reverse-flow length is reduced by about half to $\Delta x_S = 108\,\mathrm{mm}$ and χ_w to a maximum value of 61 % compared with 82 % for LFST. For HFST the mean value of c_f is always positive but χ_w has still a maximum value of 12.5 % indicating that there is instantaneous reverse flow in the near-wall region.

The reduction or elimination of weak reverse-flow regions by FST on the blades of a radial fan in one of our wind tunnels has not only reduced flow losses and the noise level but also low-frequency oscillations in the tunnel (see also [18]).

3 Strong Reverse-Flow Regions

Strong reverse-flow regions, as defined in the introduction, occur, for example, upstream and downstream of obstacles, such as a two-dimensional fence, or downstream of a backward facing step. For a specific case the separation line is generally fixed and the reattachment line of the closed reverse-flow region is the object of the manipulation. In the case of the reverse-flow region upstream of an obstacle the reattachment line is generally fixed and only the separation line can be manipulated [53]. This introduces a severe restriction in comparison with WRF regions.

Two cases (Fig. 20) will be considered briefly: Firstly, the flow over a 2-D fence with manipulation by means of a spoiler or a ZNMF actuator upstream of the fence [53, 54] and, secondly, the same flow with manipulation by a ZNMF actuator at the tip of the straight or swept fence [33].

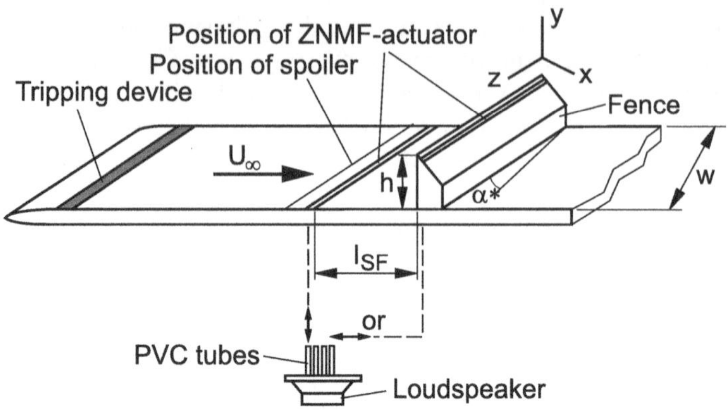

Fig. 20. Schematic view of the fence test configuration. From Huppertz & Fernholz [33]

3.1 Manipulation of the Closed Reverse-Flow Region Downstream of a Fence

Measurements downstream of a straight backward facing step were performed by Huppertz [32] and downstream of a swept step by Fernholz et al. [20] and Kaltenbach & Janke [37]. The latter paper and the one by Wengle et al. [64] compare the measured data with a direct numerical simulation. For lack of space the reader is referred to the original publications about the step flows.

The fence flows have in common that there is a well defined 2-D boundary layer upstream, the thickness of which is of the order of the fence height, h, an aspect ratio $w/h \leq 50$ and a blockage ratio $H/h \leq 41$. The upstream boundary layer was tripped and was fully turbulent. The corresponding profiles are published in Huppertz & Fernholz [33] and Siller & Fernholz [58]. The free-stream turbulence level Tu_δ is below 0.3 % and the wave form of the excitation is sinusoidal and two-dimensional in spanwise direction. The sweep angles α^* are 0° or 20°, cases with larger sweep angles $\leq 50°$ were investigated but are not discussed here for lack of space. Finally, the ratio b/h (b is the slot width) was kept within the range $0.05 \leq b/h \leq 0.12$ and the optimal non-dimensional distance between the fence and the upstream actuator l_{SF}/h was found to be between 1.5 and 2. It has been shown, for example by Siller [53], that the non-dimensional reattachment length $\overline{x}_{R_0}/h$ without forcing is mainly dependent on the blockage ratio for $Re_h \geq 5000$. Results for $\overline{x}_{R_0}/h = f(H/h)$, from the investigations of Siller [53], Good & Joubert [25], Castro & Fackrell [5] and Durst & Rastogi [13] are shown in Fig. 21. The parameters of the investigations discussed subsequently and some results of the optimization are presented in Table 1.

The main differences between these fence-flow experiments are the location of the actuator (UF or TF) and the respective optimal parameters for the non-dimensional frequency and amplitude.

Table 1. $St_{opt} = (fh/U_\infty)_{opt}$, $Re_h = hU_\infty/\nu$, $\tilde{c}_p = p_{rms}/\rho U_\infty^2$, $A = v_{rms}/U_N$, $c_{\mu_h} = \rho b v_{rms}^2/(\rho h U_\infty^2/2)$, L_p measured in the pressure chambers

Author	Actuat. Pos.	St_{opt}	Re_h	$\frac{\Delta H^*}{h}$	L_p [dB]	$\tilde{c}_p$	A	c_{μ_h}	Tu_δ [%]	$\frac{\delta}{h}$	$\frac{H}{h}$	$\frac{w}{h}$	α^*	$\frac{b}{h}$	$\left(\frac{l_{SF}}{h}\right)_{opt}$	$\frac{x_{R_0}}{h}$	$\frac{x_R}{x_{R_0}}$
Siller &	UF	0.05	10300	0.28	–	– ▽	–	–	0.3	0.8	6.75	22	0	Spoiler	3	14.1	0.67
Fernholz	UF	0.05	10700	–	130	0.80 ▲	–	–	0.3	0.8	6.75	22	0	0.05	1.8	14.1	0.74
(1997)	UF	0.055	10700	–	137	1.78 ■	–	–	0.3	0.8	9	22	0	0.05	2	14.8	0.51
Siller (1999)	UF	0.050	10500	–	137	1.92	2.1	0.441	0.3	0.8	6.75	22	0	0.05	1.75	13.6	0.60
Huppertz	UF	0.05	5330	–	–	–	0.88	0.186	0.2	2.0	41	50	0	0.12	1.5	16	0.68
&	UF	0.05	5330	–	–	–	0.88	0.186	0.2	2.0	41	≈ 50	20	0.12	1.5	15	0.69
Fernholz	TF	0.128[a]	5330	–	–	–	0.88	0.186	0.2	2.0	41	50	0	0.12	0	16	0.65
(2002)	TF	0.128[a]	5330	–	–	–	0.88	0.186	0.2	2.0	41	≈ 50	20	0.12	0	15	0.69

[a] $St_{\delta_2} = 0.018$

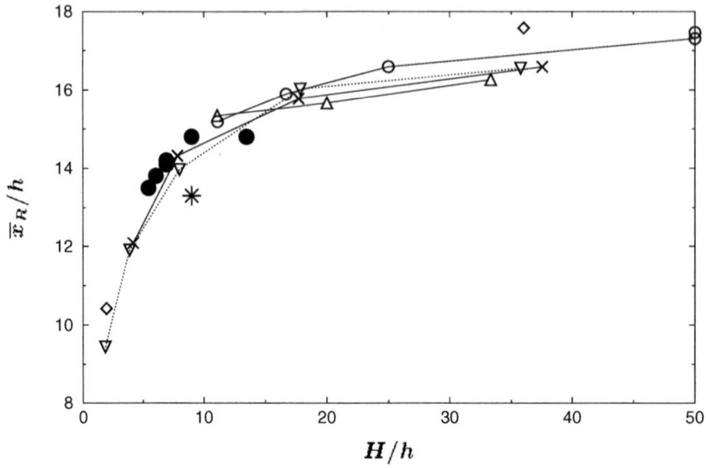

Fig. 21. The length of the reverse-flow region downstream of the fence $\overline{x}_R$ as a function of blockage. $*$ Good & Joubert [25], $\circ\triangle$ Castro & Fackrell [5], $\nabla\diamond\times$ Durst & Rastogi [13], $\bullet$ Siller [53]. From Siller [53]

In case UF the upstream and the downstream reverse-flow regions must be manipulated in order to reduce the length of the downstream separation bubble. In essence this means that upstream of the fence a roller with the diameter of about h must be generated to lead the separated shear layer over the fence and to change the flow angle in the downstream direction above the fence, and further that large coherent structures must originate in the shear layer close downstream of the second separation at the tip of the fence. These coherent structures are caused by a stochastic roll-up process due to the inflectional velocity profiles. On average there are preferred frequencies (Fig. 22) which show as a bump in the power spectral density distribution of the streamwise velocity and which shift to higher frequencies with increasing mean velocity [33].

The frequencies of these amplified waves have been shown to be effective excitation frequencies and lie here in the range of 40 to 50 Hz, giving a non-dimensional frequency of $St_h \approx 0.05$. This agrees with the result of Siller [53] and lies within the band of the shedding frequency of the bubble downstream of the fence. With this Strouhal number (Fig. 23(a)) the non-dimensional reattachment length $\overline{x}_R/\overline{x}_{R_0}$ (where $\overline{x}_{R_0}$ is the distance between the fence and the reattachment line without manipulation) was determined as a function of the excitation amplitude $A = v_{rms}/U_N$ at the slot exit (U_N is the velocity in the free-stream normal to the fence which is the characteristic quantity to account for various sweep angles α^*). For further investigations A was chosen as 88 % because this is the value of A where a change in slope occurs towards an onset to saturation. With this value of A, $\overline{x}_R/\overline{x}_{R_0}$ was determined as a function of the excitation frequency (Fig. 23(b)) and the curves for sweep

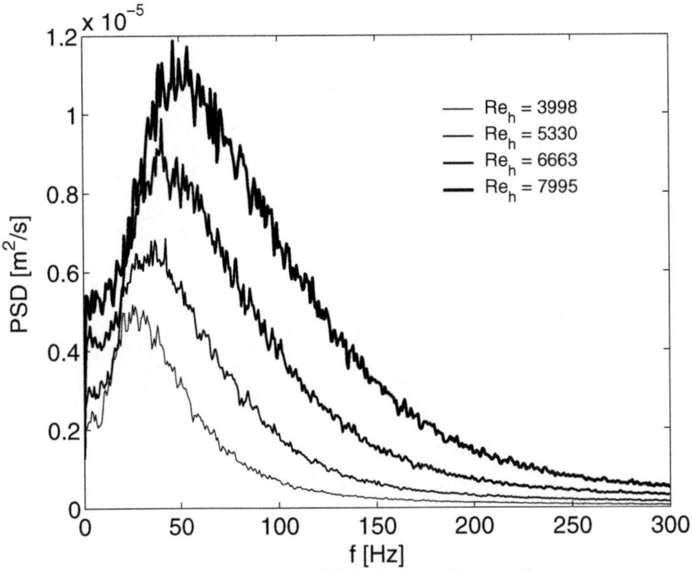

Fig. 22. Power spectral density distribution at various free-stream velocities U_N ($x/h = 10$, $y/h = 3.8$, no manipulation). From Huppertz & Fernholz [33]

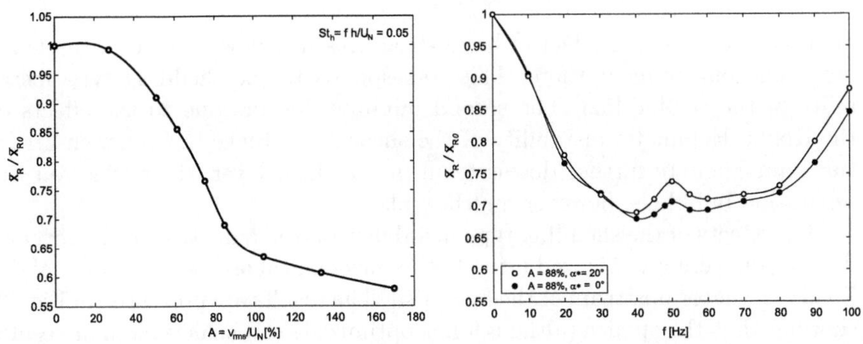

(a) (b)

Fig. 23. Non-dimensional reverse-flow length $\overline{x}_R/\overline{x}_{R_0}$ for the excitation upstream of the fence (23(a)) as a function of the excitation amplitude ($St_h = 0.05$, $\alpha^* = 20°$) and (23(b)) as a function of the excitation frequency, and the sweep angle α^*. From Huppertz & Fernholz [33]

angles 0° and 20° confirm the above optimal Strouhal-number. The influence of sweep is almost negligible in this case.

Siller [53] extended the frequency range to higher values using the sound-pressure coefficient as a parameter for the amplitude (Fig. 24) and found two local minima (see also [9], for a backward facing step flow). The first mini-

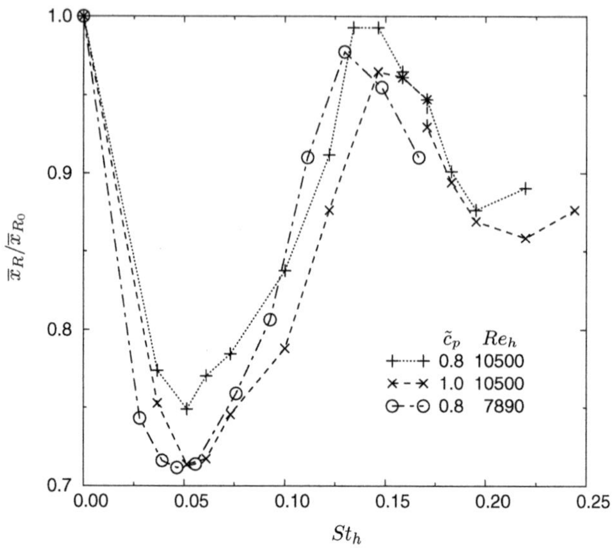

Fig. 24. Non-dimensional reverse-flow length $\overline{x}_R/\overline{x}_{R_0}$ as a function of St_h for two different amplitudes and a constant distance of the slot to the fence $l_{SF} = 1.75h$ ($l_{SF} = 2.0h$ for $Re_h = 7890$). From Siller [53]

mum is caused by the effect of large structures just downstream of the fence increasing entrainment there. This corresponds to the shedding type instability of the bubble [52]. The second minimum occurs due to the effects of the Kelvin-Helmholtz instability of the shear layer. Since the coherent structures amalgamate further downstream in the shear layer, their effect on the separation bubble is, however, much weaker.

The effects of the shedding type instability on the reduction of the reverse-flow region were investigated using both a mechanical oscillating spoiler and a ZNMF-actuator upstream of the fence [54]. The results are presented in Fig. 25 showing that the spoiler (although not optimized) can achieve similar results as the ZNMF-actuator which, however, is more versatile and mechanically simpler. $\Delta H^*/h$ is the dimensionless height of the spoiler above the wall, which was here 28 % of the fence height, and l_{SF}/h the location of the actuator upstream of the fence (see also Table 1). The optimum value for l_{SF}/h (spoiler) was 3 and about 2 for the ZNMF-actuator.

Figure 26 shows the wall values of the flow downstream of the fence for an optimized ZNMF-actuator (case UF) taken from Siller [53]. The distributions of the pressure coefficient $c_p = 2(p(x) - p_{ref})/(\rho U_\infty^2)$, the mean wall shear-stress coefficient $c_f = 2\overline{\tau}_w/(\rho U_\infty^2)$, the reverse-flow factor χ_w, and the rms fluctuation value $c_f' = 2(\overline{\tau_w'^2})^{1/2}/(\rho U_\infty^2)$ are plotted against $(x - x_f)/\overline{x}_R$. Here $\overline{x}_R$ is the time averaged length of the reverse-flow region. In this scaling (see also [46]) the distributions of c_p and c_f, respectively, fall onto each other with and without forcing. The distributions of χ_w and c_f' in the bubble are very

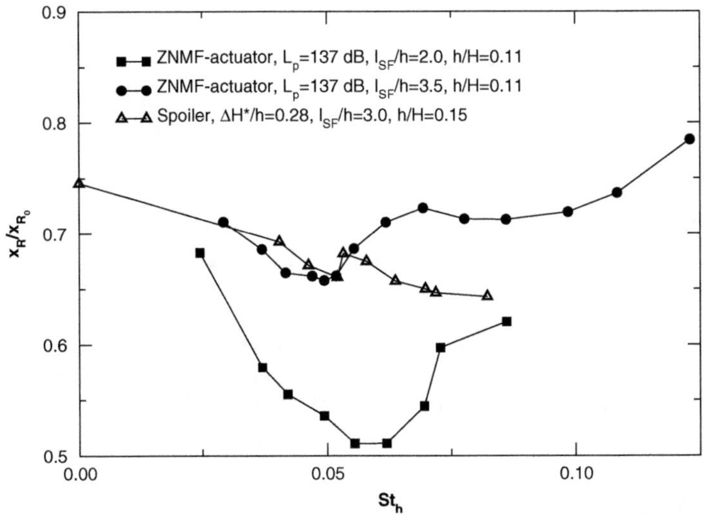

Fig. 25. Non-dimensional reverse-flow length as a function of St_h for an oscillating spoiler and a ZNMF-actuator with different blockage ratio and actuator position. From Siller & Fernholz [54]

much affected by the high forcing amplitude and do not collapse. The large increase of c_f' due to forcing is also reflected in the distributions of the higher moments of c_f' with the skewness increasing and the flatness decreasing (see [53]).

Manipulating the downstream reverse-flow-region with an actuator at the tip of the fence (TF) appears to have advantages, at least at first sight. We shall find that the necessary excitation amplitudes can be smaller but that the frequencies must be higher, with the overall result that the reduction of the bubble length is not much different from that of case UF. The main mechanism for the manipulation is the Kelvin-Helmholtz instability using single and bimodal forcing [33]. There the appropriate local Strouhal number is $St_{\delta_2} = f\delta_2/U_{\max}$, where δ_2 is the momentum-loss thickness of the separated shear layer and $U_{\max}(x)$ the maximum velocity of the profile. St_{δ_2} was determined as 0.020 resulting in an excitation frequency $f = 114\,\text{Hz}$. The results from the parameter optimization are shown in Fig. 27. Here the non-dimensional reattachment length $\overline{x}_R/\overline{x}_{R_0}$ is plotted against the excitation frequency as a function of two sweep angles and two amplitudes $A = v_{rms}/U_N$. The optimum excitation frequency is approximately 100 Hz, independent of sweep angle and amplitude, and close to the maximum excitation frequency of the Kelvin-Helmholtz instability. With the optimum Strouhal number $St_{\delta_2} = 0.018$ ($St_h = 0.128$) the development of the reattachment length as a function of the amplitude v_{rms}/U_N shows a minimum at $A = 85\,\%$ (for $\alpha^* = 0°$ this value is practically identical). Huppertz & Fern-

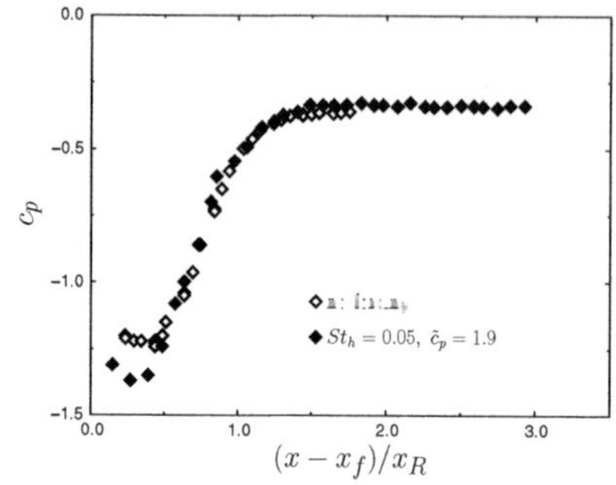

(a) Distributions of the pressure coefficient c_p downstream of the fence with and without manipulation (from [53]).

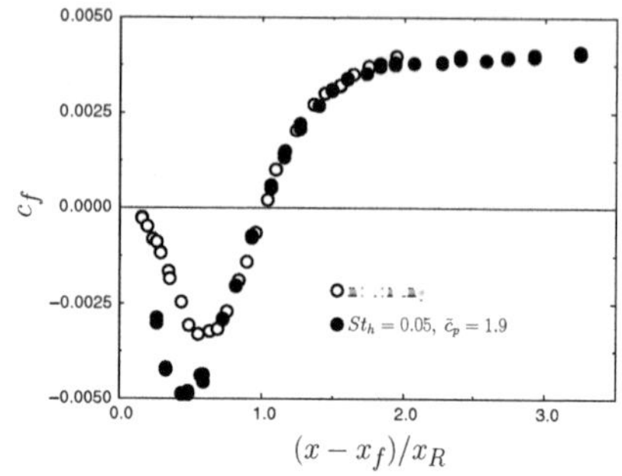

(b) Distributions of the mean skin-friction coefficient c_f downstream of the fence with and without manipulation (from [53]).

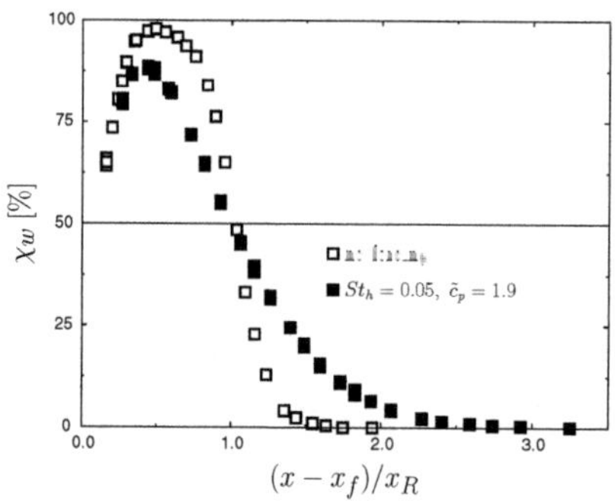

(c) Distributions of the reverse-flow factor χ_w downstream of the fence with and without manipulation (from [53]).

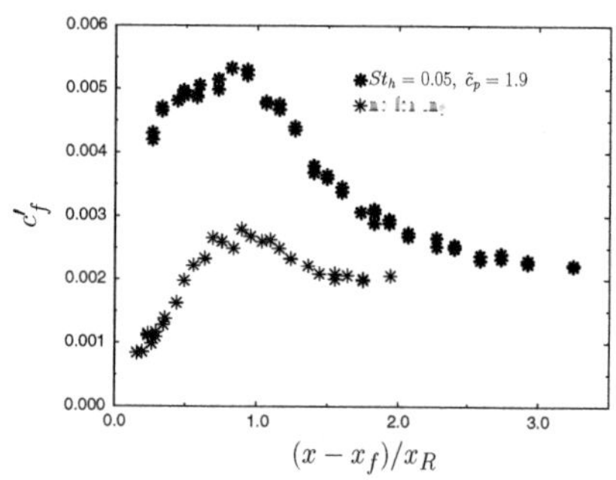

(d) Distributions of the fluctuating skin-friction coefficient c'_f downstream of the fence with and without manipulation (from [53]).

Fig. 26.

holz [33] have explained why the excitation of the Kelvin-Helmholtz instability is optimal at $f = 114\,\mathrm{Hz}$ taking into account the behavior of the kinetic energy of the coherent structures E_{coh} integrated over the width of the shear layer:

$$E_{coh}(u, f_0) = \int_0^\infty \frac{\overline{u'^2_{coh}}(f_0)}{U_N^2}\, d(y/h) \quad \text{with} \quad \overline{u'^2_{coh}}(f_0) = \int_{f_0-\Delta f}^{f_0+\Delta f} S_{uu}(f)\, df$$

where S_{uu} is the power spectral density of the u' fluctuation (note that: $\overline{u'^2} = \int_0^\infty S_{uu}(f)\, df$), $[f_0 - \Delta f, f_0 + \Delta f]$ is the spectral bandwidth of the coherent energy and f_0 is the excitation frequency f_{exc} or the corresponding subharmonic $f_{sub} = f_{exc}/2$. No subharmonic component of f_{exc} could be detected in the frequency spectrum of the shear layer, i.e. no vortex pairing

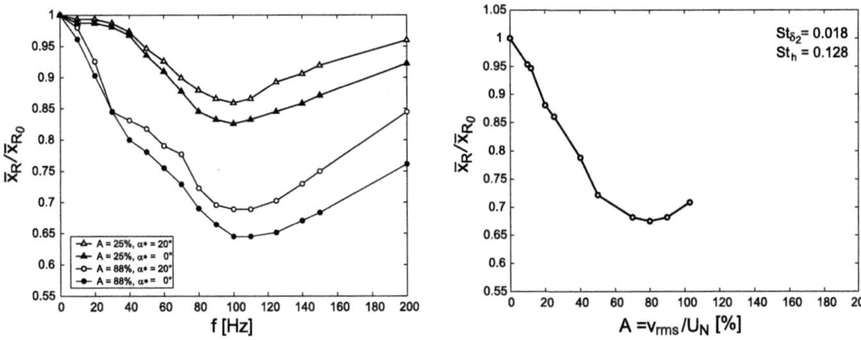

(a) (b)

Fig. 27. Non-dimensional reverse-flow length $\overline{x}_R/\overline{x}_{R_0}$ for the excitation at the tip of the fence (27(a)) as a function of the excitation frequency and the sweep angle α^* and (27(b)) as a function of the excitation amplitude ($St_{\delta_2} = 0.018$, $\alpha^* = 20°$). From Huppertz & Fernholz [33]

occurred [33]. Hence the coherent energy was only calculated for the excitation frequency for $A = 0.25$.

Figure 28 presents the development of $E_{coh}(u)$ in streamwise direction for $\alpha^* = 20°$ and $A = 25\,\%$ at four excitation frequencies ($f = \{51, 75, 102, 150\}$ Hz) and a bandwidth of $\Delta f = 4$ Hz. Amplification, saturation and a decrease of the coherent energy can be observed. Each curve has a saturation value $E_{coh_{\max}}$ at a different height and at a different streamwise location x/h. This behavior is typical of a Kelvin-Helmholtz instability. A comparison of Fig. 28 and 27(a) shows that the largest reduction of the reverse-flow region ($\overline{x}_R/\overline{x}_{R_0} = 0.84$) occurs at $f = 120$ Hz which is not the curve with the absolute maximum ($f = 51$ Hz) but the curve with the highest values of E_{coh} very close to the fence, since here the flow manipulation is most effective in reducing the reattachment length.

For this case and single frequency excitation, PIV measurements were made which show the expected roll-up of the shear layer to a large coherent structure just downstream of the fence without vortex pairing (not shown here, but see [33]).

After the discussion of single frequency excitation we shall now briefly refer to bimodal forcing. Many investigations of free shear layers have shown that the control range of forcing can be extended by adding the first subharmonic frequency $f_{sub} = f_{opt}/2$ (e.g. [34]). The excitation signal is then described by

$$\text{Signal} = A_{fun} \sin(2\pi f t) + A_{sub} \sin(2\pi f t/2 + \Delta\varphi) \quad \text{with} \quad 0° \leq \Delta\varphi < 180°.$$

For the manipulation of the shear layer by bimodal excitation the amplitudes and the frequencies of the fundamental and the subharmonicexcitation and the initial phase difference $\Delta\varphi$ must be determined. The optimization process

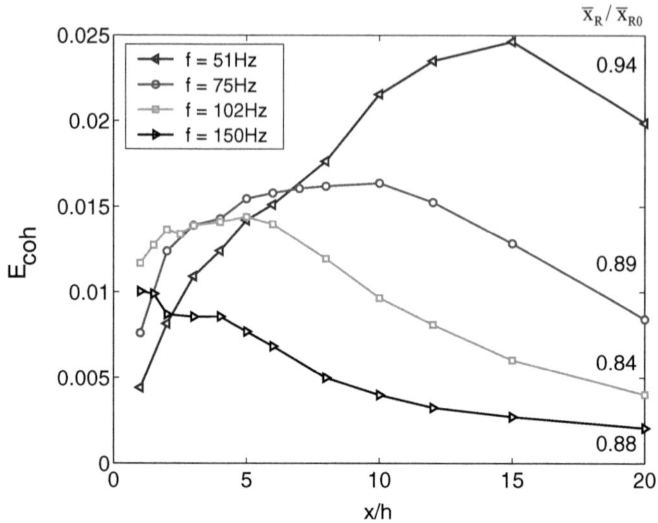

Fig. 28. Development of the coherent kinetic energy $E_{coh}(u)$ at four excitation frequencies for $A = 25\,\%$, $\alpha^* = 20°$ and $Re_h = 5330$. From Huppertz & Fernholz [33]

has been described in detail by Huppertz & Fernholz [33]. In order to obtain an efficient reduction of the reattachment length, the energy containing structures must be generated further upstream than they would be generated with $f_{fun} = 102\,\text{Hz}$ and $f_{sub} = 51\,\text{Hz}$ and this was achieved by choosing $f_{fun} = 204\,\text{Hz}$, $f_{sub} = 102\,\text{Hz}$ and $\Delta\varphi = 80°$. In this case vortex pairing is forced at about $x/h = 5$ leading to increased entrainment and a reduction of $\overline{x}_R/\overline{x}_{R_0}$ of 25 %.

The results of single mode and bimodal excitation on the development of the wall parameters downstream of the fence are shown in Fig. 29 where the abscissa is $x/\overline{x}_R$, with $x = 0$ at the fence. As in Fig. 26 the distributions of c_p and c_f, additionally that of χ_w, are quasi self-similar. The exception is the one case where the excitation was introduced upstream of the fence (UF), a case comparable to that of Siller [53]. The distribution of χ_w shows that instantaneous reverse flow occurs to about one bubble length downstream of mean reattachment. With increasing excitation amplitude the peak values of c_f' increase ($\approx 100\,\%$) and move upstream of the mean reattachment line.

3.2 Manipulation of Strong Reverse Flow Caused by a Trapped Vortex

Here we present two cases of strong reverse-flow regions which are caused by pressure induced separation where the large negative skin friction was generated by a "trapped" vortex.

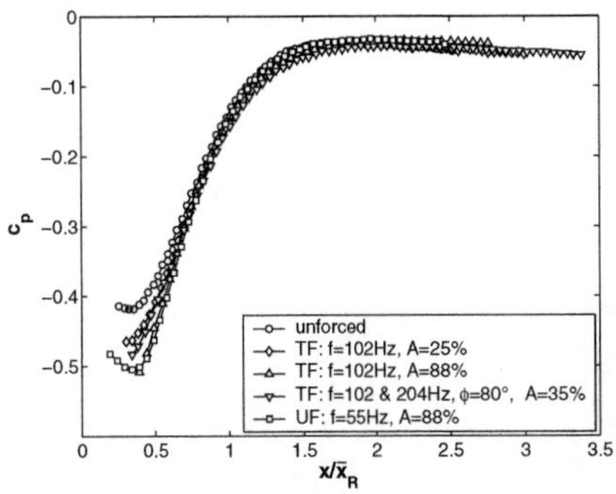

(a) Distributions of the pressure coefficient c_p downstream of the fence ($\alpha^* = 20°$) with and without manipulation (see [33]).

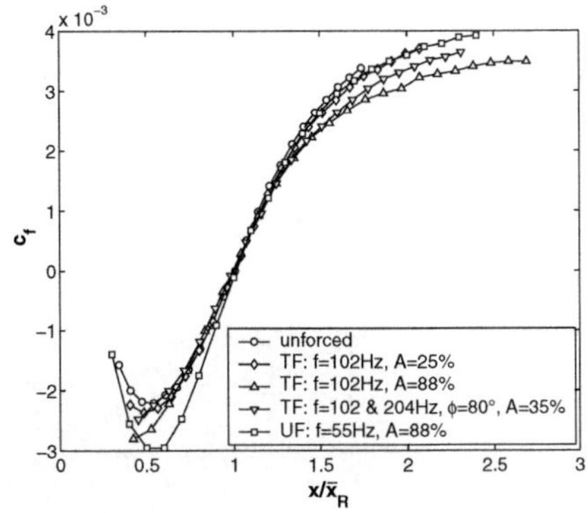

(b) Distributions of the mean skin-friction coefficient c_f downstream of the fence ($\alpha^* = 20°$) with and without manipulation (see [33]).

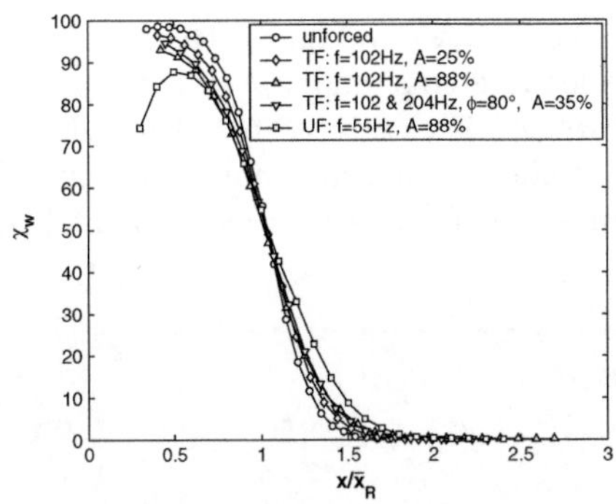

(c) Distributions of the reverse-flow factor χ_w downstream of the fence ($\alpha^* = 20°$) with and without manipulation (see [33]).

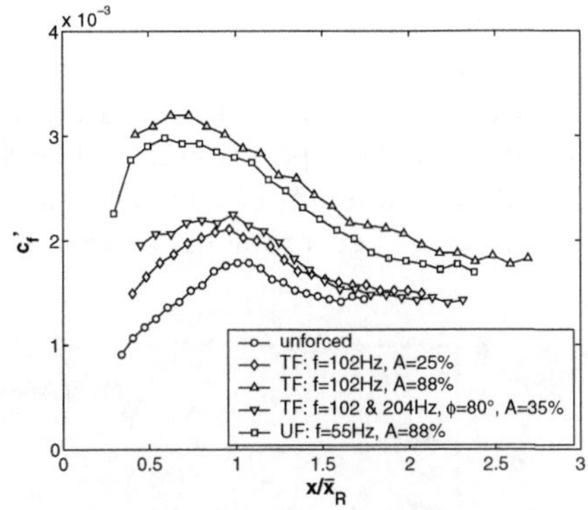

(d) Distributions of the fluctuating skin-friction coefficient c_f' downstream of the fence ($\alpha^* = 20°$) with and without manipulation (see [33]).

Fig. 29.

In the first case the vortex was in front of a two-dimensional obstacle emerging above a turbulent boundary layer. Figure 30 combines a flow visualization [65] and a skin friction distribution measured by a surface fence [45]. Such a vortex and its high skin friction is responsible for scouring out the bed of a stream in front of a bridge pillar, for example.

A similar flow, this time over a two-dimensional fence, was investigated by Siller & Fernholz [58] without manipulation and by Huppertz & Fernholz [33] with manipulation. In the former case the distributions of the characteristic mean wall values upstream of the fence c_p, c_f and χ_w as a function of the

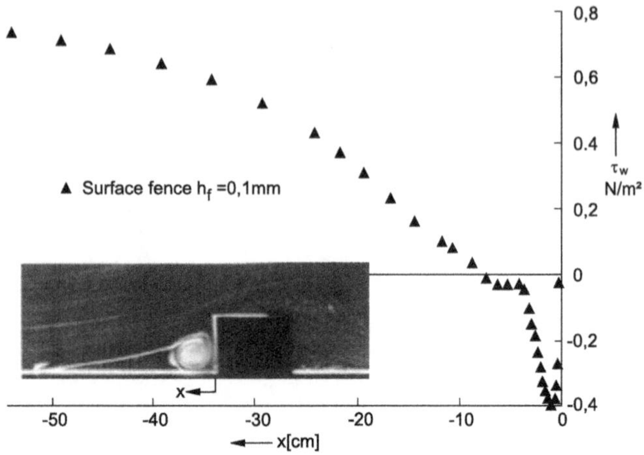

Fig. 30. Flow visualization [65] and wall shear-stress distribution [45] upstream of a two-dimensional obstacle in a turbulent boundary layer

non-dimensional length $(x_f - x)/(x_f - x_D)$ are shown in Fig. 31. Here x_f is the position of the fence measured from the leading edge and $x_f - x_D$ the mean position of the separation line. The SRF-region is clearly visible although the negative maximum of the skin friction coefficient c_f could not be measured owing to design problems of the configuration. In a related experiment [33]

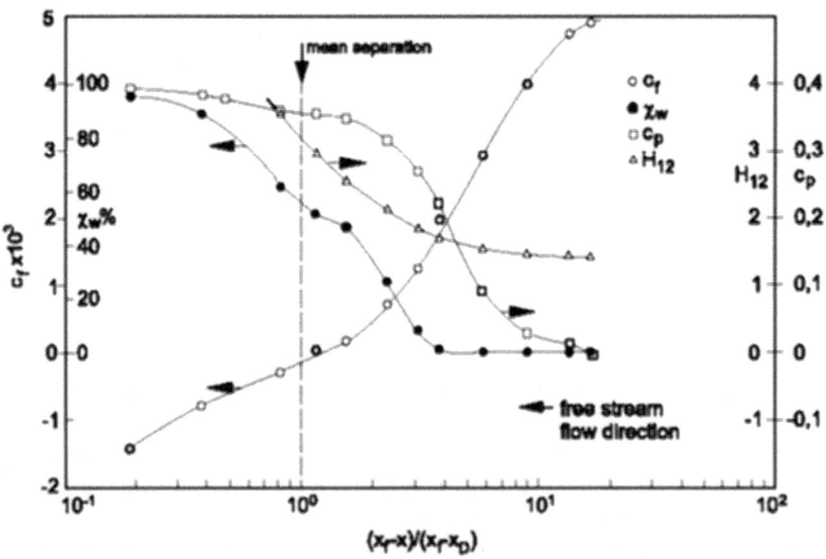

Fig. 31. Streamwise distribution of the static pressure coefficient c_p, the skin-friction coefficient c_f, the reverse-flow parameter χ_w, and the shape parameter H_{12} (lines are for visual aid only). From Siller & Fernholz [58]

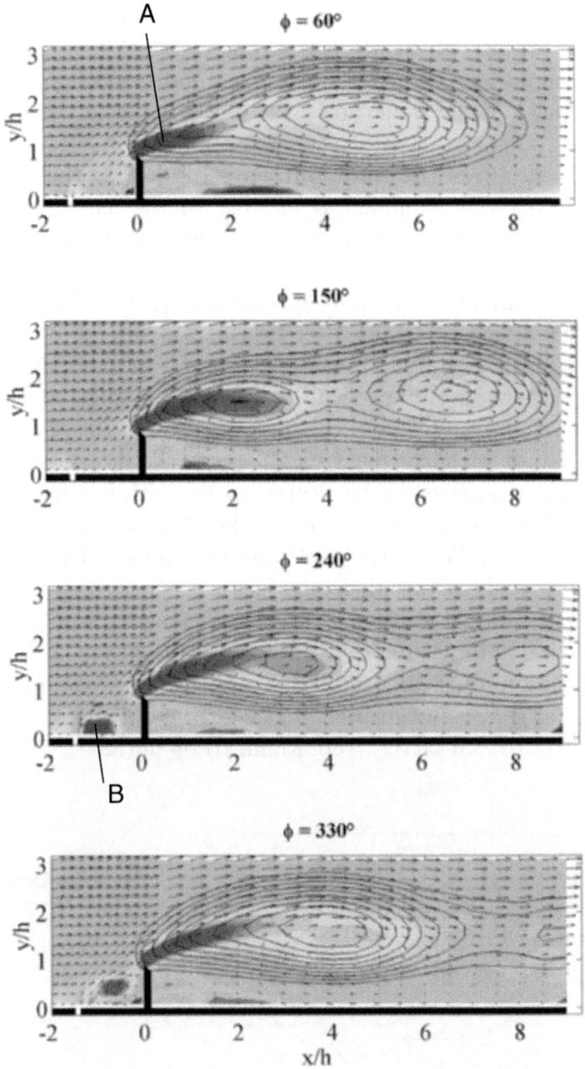

Fig. 32. Phase-averaged vorticity component $\langle\omega_z\rangle$ at $St_h = 0.07$, $A = 88\,\%$, and $\alpha^* = 20°$ (excitation upstream of the fence) for four phase angles. From Huppertz & Fernholz [33]

the flow over the fence was manipulated by a ZNMF-actuator upstream of the fence in order to reduce the length of the reverse-flow region downstream of the fence. The manipulation of the trapped vortex in front of the fence is demonstrated by means of PIV measurements of the phase-averaged vorticity and the streamlines. Figure 32 shows the vorticity component $\langle\omega_z\rangle =$

$\langle \partial \bar{v}/\partial x \rangle - \langle \partial \bar{u}/\partial y \rangle$ in the xy-plane $-2 \leq x/h \leq 8$ and $0 \leq y/h \leq 3$ at four phase angles φ (at $A = 88\,\%$, $St_h = 0.07$ and $\alpha^* = 20°$).

The streamlines are viewed from a coordinate system moving with the convection velocity of the structures $\bar{u}_c \approx 0.6 U_N$. During the "suction" phase of the actuator (φ approximately between 30° and 150°) a structure (A) forms downstream of the tip of the fence. Subsequently during the "blowing" phase the trapped vortex (B) is enhanced in front of the fence, the angle of the streamlines at the fence tip changes and structure (A) is convected downstream.

In a second, completely different experiment ([21], and [70] and 1993) the vortex was generated and trapped on the suction side of a wedge-like body which consisted of a NACA 0025 wing section (1 m chord length) with the sharp trailing edge pointing upstream (Fig. 33). The angle of incidence could be varied between 0° and 27° and the manipulation of the separated flow was achieved by a flapping sharp-edged spoiler (note the direction) at the leading edge. The spoiler was driven by an electric motor with frequencies up to 70 Hz and an amplitude between 0.5 and 2 mm. The chord Reynolds number was in the range $3.7 \times 10^5 \leq Re_c \leq 1.4 \times 10^6$ and two angles of incidence, 18° and 27°, were investigated because their separation scenarios and the results of the manipulation, measured as lift and drag coefficients c_L and c_D, were rather different. For the manipulation, the appropriate amplitude Δh^* and frequency f had to be chosen. With an amplitude $\Delta H^* = 1$ mm the non-dimensional frequency $St_c = fc/U_\infty$ was varied between 0 and 14 (Fig. 34) and values between 1 and 2 (10 and 20 Hz) were found to be optimal.

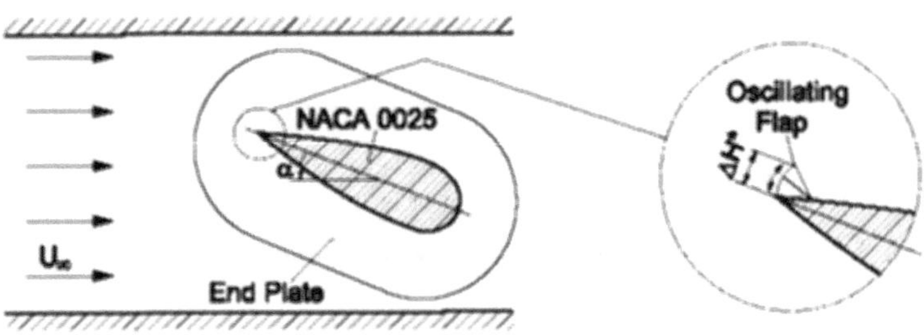

Fig. 33. Test configuration. From Fernholz et al. [21]

At $\alpha = 18°$ the skin-friction distribution (Fig. 35) showed that a closed reverse-flow region existed on the suction side between $0.10 \leq x/c \leq 0.52$ which was reduced in length by the manipulation and moved upstream ($x_R/c = 0.36$). In both cases the reverse-flow region is strong, as shown by the high negative values of the skin friction which were measured by means of a wall pulsed-wire probe. With manipulation the pressure distribution has a higher suction peak but the overall distribution of c_p produced slightly less lift than without manipulation ($c_L = 0.84$ against 0.88 measured by a

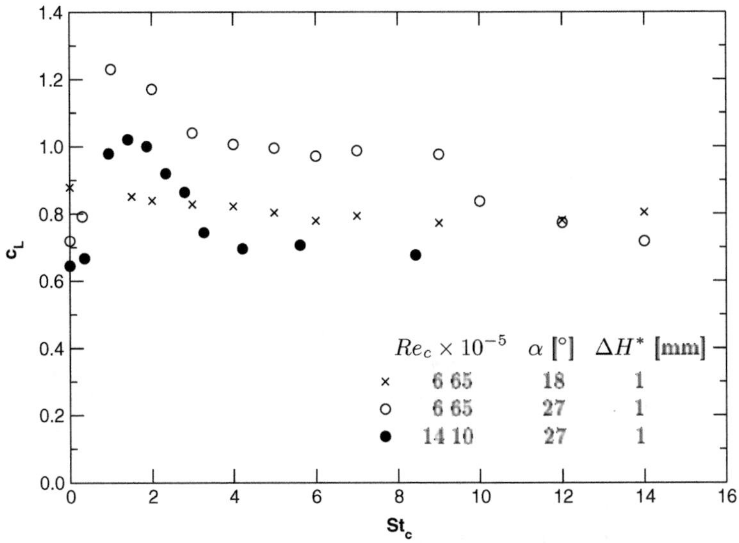

Fig. 34. Variation of the lift coefficient c_L as a function of chord Strouhal number, Reynolds number Re_c, angle of incidence α at a fixed spoiler amplitude $\Delta H^* =$ 1 mm. From Zhou et al. [70]

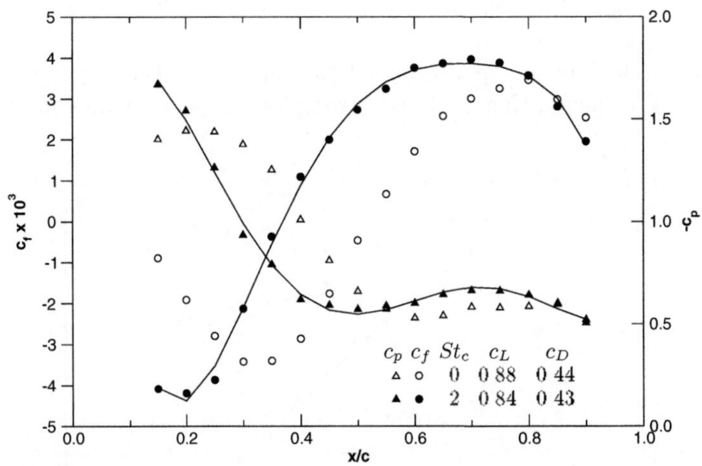

Fig. 35. Distribution of pressure coefficient c_p and skin-friction coefficient c_f with ($St_c = 2$) and without ($St_c = 0$) manipulation ($Re_c = 6.7 \times 10^5$, $\alpha = 18°$, $\Delta H^* =$ 1 mm, lines are for visual aid only). From Zhou et al. [70]

six-component mechanical balance). The drag coefficient was practically unchanged.

At $\alpha = 27°$, the skin friction distribution (Fig. 36) shows a large open separation region (post stall) with weak or no reverse flow in the upstream

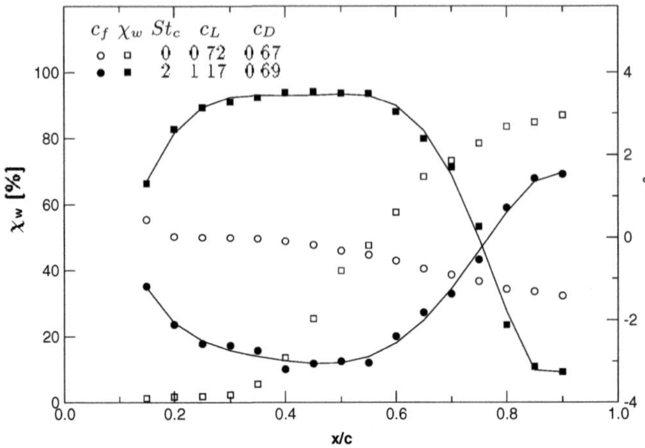

Fig. 36. Distribution of mean skin-friction coefficient c_f and reverse-flow factor χ_w with and without manipulation ($Re_c = 6.5 \times 10^5$, $\alpha = 27°$, $\Delta H^* = 1$ mm, lines are for visual aid only). From Zhou et al. [70]

part. Both c_f and χ_w are practically zero ("dead air region") and reach higher values only towards the rear of the wedge-like body owing to the massive open separation. The pressure distribution remained flat ($c_p \approx -0.85$) over the greater part of the wing (Fig. 37).

Manipulation by the spoiler forced the separated shear layer to reattach ($x_R/c \approx 0.77$) generating a closed strong reverse-flow region in the range

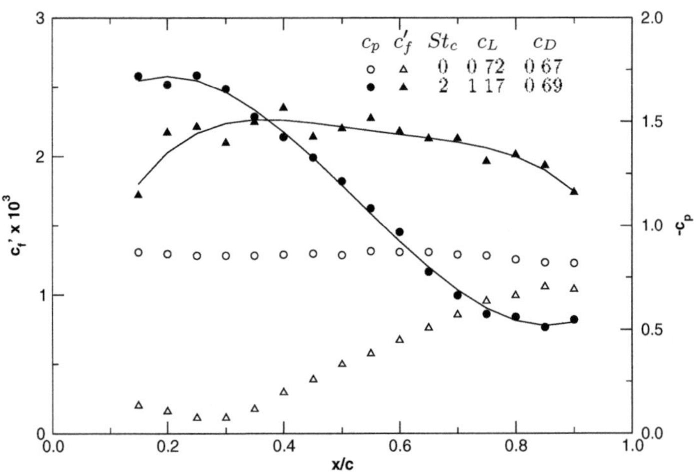

Fig. 37. Distributions of pressure coefficient c_p and fluctuating skin-friction coefficient c_f' with and without manipulation ($Re_c = 6.5 \times 10^5$, $\alpha = 27°$, $\Delta H^* = 1$ mm, lines are for visual aid only). From Zhou et al. [70]

$0.10 \leq x/c \leq 0.77$ as shown by the distributions of the skin friction and reverse flow parameter (Fig. 36). This picture reveals a similar behavior of the flow to what was found for $\alpha = 18°$. χ_w increased to about 90 % and c_f reached values of about -3×10^{-3} on the upstream half of the suction side indicating forward flow towards the rear end of the wedge-like body. The pressure distribution (Fig. 37) shows a large increase of the suction pressure resulting in a rise of c_L from 0.72 to 1.17, again with c_D remaining almost constant. Measurements of the wake profiles showed however, that c_D decreased with manipulation. All these changes of relevant flow parameters point to a large vortex with fairly stable circulation which occupied more than half of the suction side. The effect of the vortex near the wall is also reflected in the increase of the fluctuating skin friction coefficient c'_f the distribution of which is shown in Fig. 37.

From this we may conclude that (a) it can be detrimental if the reverse-flow region is shortened but moved at the same time upstream and into the region of the suction peak (case 18°) and (b) that if open separation is converted to closed separation (case 27°) the lift may increase considerably due to the strong vortex captured in the bubble (see also [71], and [67]).

4 Conclusions

With growing experience from one test case to the next a certain pattern evolved as to how to approach the successful manipulation of wall-bounded turbulent shear-flows: Clearly define the goal of the manipulation, find the flow region most suitable for the manipulation, decide whether the actuator should be passive or active, determine the optimum properties of the actuator signal and finally choose the quantity by which the success of the manipulation can best be controlled.

In the case studies presented here, manipulation mainly had the goal of reducing the length of the separation region as much as possible or increasing the lift in the post-stall regime on an airfoil. Flows over wings with large separation regions show interesting physical phenomena if manipulated, depending on whether it is leading-edge or trailing-edge stall. The results are often specific to an airfoil type and difficult to generalize, however.

The effectiveness of an actuator is severely reduced if it is not located upstream and close to the mean separation line. Actuators built into the wall have the advantage of not generating additional drag when they are not in operation and therefore mainly active zero net-mass flux actuators consisting of a piston (loudspeaker), a pressure chamber and a spanwise slot were used. In three cases mechanical spoilers were employed which worked almost equally well but are less versatile than the ZNMF-actuators and can be run only at frequencies below 50 Hz. The ZNMF-actuators function at low and high frequencies, can vary the phase and the amplitude of the signal over a wide range and generate, if two-dimensional, spanwise vortices and, if short and driven in antiphase, longitudinal vortices. This gives them a wide range of applications, especially the possibility of reducing the unsteadiness of the separation region.

In all case studies presented here, the skin friction was the quantity which was the best indicator of the state of the separation region and the strength of the reverse flow. Since in many cases the measuring time can be kept very short skin friction may be used as a convenient control variable for closed-loop control of separation regions.

Acknowledgment

The authors acknowledge the advice and the assistance of P. Erk, M. Goetz, A. Huppertz, M. Kalter, H. A. Siller, and M. D. Zhou whose articles have largely been used to write this survey paper. Thanks are also due to P. J. Finley for critical comments on the manuscript.

References

[1] Ahuja, K. K. & Burrin, R. H. 1984 Control of flow separation by sound. *AIAA Paper 84–2298*.

[2] Alving, A. E. & Fernholz, H.-H. 1996 Turbulence measurements around a mild separation bubble and downstream of reattachment. *J. Fluid Mech.* **322**, 297–328.

[3] Amitay, M. & Glezer, A. 2002 Role of actuation frequency in controlled flow reattachment over a stalled airfoil. *AIAA J.* **40**, 209–216.

[4] Béraud, F. 1994 Experimental study of the receptivity of a separated boundary layer on an airfoil at high angles of attack and low Reynolds numbers. Diplomarbeit, Technische Universität Berlin, Hermann-Föttinger-Institut für Strömungsmechanik.

[5] Castro, I. P. & Fackrell, J. E. 1978 A note on two-dimensional fence flows with emphasis on wall constraints. *J. Indust. Aerodyn.* **3**, 1–20.

[6] Chen, F.-J., Yao, C., Beelev, G. B., Bryant, R. G. & Fox, R. L. 2000 Development of synthetic jet actuators for active flow control at NASA Langley. *AIAA Paper 2000–2405*.

[7] Cheng, X. 1991 Experimental research on control of flow separation by means of excitation. Ph. D. thesis, Nanjing Aeronaut. Inst., Nanjing, P. R. China.

[8] Cherry, N. J., Hillier, R. & Latour, M. E. M. P. 1984 Unsteady measurements in a separated and reattaching flow. *J. Fluid Mech.* **144**, 13–46.

[9] Chun, K. B. & Sung, H. J. 1996 Control of turbulent separated flow over a backward facing step by local forcing. *Exp. in Fluids* **21**, 417–426.

[10] Coller, B. D., Noack, B. R., Narayanan, S., Banaszuk, A. & Khibnik, A. I. 2000 Reduced-basis model for active separation control in a planar diffuser flow. *AIAA Paper 2000–2562*.

[11] Compton, D. A. & Johnston, J. P. 1992 Streamwise vortex production by pitched and skewed jets in a turbulent boundary layer. *AIAA J.* **30** (3), 640–647.

[12] Dengel, P. & Fernholz, H.-H. 1990 An experimental investigation of an incompressible turbulent boundary layer in the vicinity of separation. *J. Fluid Mech.* **212**, 615–636.

[13] Durst, F. & Rastogi, A. K. 1980 Turbulent flow over two-dimensional fences. In *Turbulent Shear Flows II* (ed. B. et al.), pp. 218–232. Springer Verlag.

[14] Erk, P., Béraud, F. & Graichen, K. 1996 Investigations in the near field of a nominally two-dimensional, strong acoustic perturbation source. In *Advances in Turbulence VI* (ed. S. Gavrilakis, L. Machiels & P. A. Monkewitz), *Fluid Mechanics and its Applications*, vol. 36, pp. 177–178. Kluwer Acad. Publ.

[15] Erk, P. & Fernholz, H.-H. 1994 An experimental study of the receptivity of a separated boundary layer on a low-Reynolds number airfoil at high angles of attack. In *Book of Abstracts Euromech Colloquium 328*. Technische Universität Berlin.

[16] Erk, P., Graichen, K. & Béraud, F. 1996 Separation control by strong acoustic forcing. In *19th International Congress on Theoretical and Applied Mechanics*. IUTAM, Kyoto, JP.

[17] Erk, P. P. 1997 *Separation Control on a Post-Stall Airfoil Using Acoustically Generated Perturbations*. *Fortschr.-Ber. VDI Reihe 7 Strömungstechnik* 328. Düsseldorf: VDI Verlag.

[18] Fehse, K.-R. 1997 Experimentelle Untersuchung zur Entstehung tieffrequenter Druckschwankungen bei Radialventilatoren. Dissertation, Technische Universität Berlin, D 83.

[19] Fernholz, H.-H. 1994 Near-wall phenomena in turbulent separated flows. *Acta Mechanica Suppl.* **4**, 57–67.

[20] Fernholz, H.-H., Janke, G., Kalter, M. & Schober, M. 1993 On the separated flow behind a swept backward facing step. In *Physics of Separated Flows - Numerical, Experimental and Theoretical Aspects* (ed. K. Gersten), pp. 200–207. Vieweg Verlag.

[21] Fernholz, H.-H., Zhou, M. D., Cheng, X. & Dengel, P. 1992 The manipulation of open and closed reverse-flow regions on a wedge-like body. In *Turbulente Strömungen in Forschung und Praxis* (ed. A. Leder), pp. 123–130. Verlag Shaker.

[22] Fiedler, H. E. & Fernholz, H.-H. 1990 On management and control of turbulent shear flows. *Prog. Aerospace Sci.* **27**, 305–387.

[23] Flynn, K. P. & Panton, R. L. 1990 The interaction of Helmholtz resonators in a row when excited by a turbulent boundary layer. *J. Acoust. Soc. Am.* **87** (4), 1482–1488.

[24] Gad-el-Hak, M. & Bushnell, D. B. 1991 Separation control: Review. *J. Fluids Eng.* **113**, 5–30, Transactions of the ASME.

[25] Good, M. C. & Joubert, P. N. 1968 The form drag of two-dimensional bluff-plates immersed in turbulent boundary layers. *J. Fluid Mech.* **31** (3), 547–582.

[26] Greenblatt, D. & Wygnanski, I. J. 2000 The control of flow separation by periodic excitation. *Prog. Aerospace Sci.* **36**, 487–545.

[27] Hillier, R. & Cherry, N. J. 1981 The effects of stream turbulence on separation bubbles. *J. Wind Eng. Ind. Aero.* **8**, 49–58.

[28] Horstmann, K.-H. & Quast, A. 1981 Widerstandsvermeidung durch Blasturbulatoren. *Tech. Rep.* FB 81–33. DFVLR.

[29] Hsiao, F.-B., Liu, C. F. & Shyu, J. Y. 1990 Control of wall-separated flow by internal acoustic excitation. *AIAA J.* **28**, 1440–1446.

[30] HUANG, L. S., BRYANT, T. D. & MAESTRELLO, L. 1988 The effect of acoustic forcing on trailing edge separation and near wake development of an airfoil. *AIAA Paper 88–3531-CP.*

[31] HUANG, L. S., MAESTRELLO, L. & BRYANT, T. D. 1987 Separation control over an airfoil at high angles of attack by sound emanating from the surface. *AIAA Paper 87–1261.*

[32] HUPPERTZ, A. 2001 Aktive Beeinflussung der Strömung stromab einer rückwärtsgewandten Stufe. Dissertation, Technische Universität Berlin.

[33] HUPPERTZ, A. & FERNHOLZ, H.-H. 2002 Active control of the turbulent flow over a swept fence. *Eur. J. Mech. B/Fluids* **21**, 429–446.

[34] HUSAIN, H. & HUSSAIN, A. K. 1995 Experiments on subharmonic resonance in a shear layer. *J. Fluid Mech.* **304**, 343–372.

[35] ISOMOTO, K. & HONAMI, S. 1989 The effect of inlet turbulence intensity on the reattachment process over a backward facing step. *J. Fluids Eng.* **111**, 87–92, Transactions of the ASME.

[36] JOHNSTON, J. P. & NISHI, M. 1990 Vortex generator jets - means for flow separation control. *AIAA J.* **28** (6), 989–994.

[37] KALTENBACH, H. & JANKE, G. 2000 Direct numerical simulation of flow separation behind a swept rearward-facing step at $Re_h = 3000$. *Phys. Fluids* **12**, 2320–2337.

[38] KALTER, M. & FERNHOLZ, H.-H. 2001 The reduction and elimination of a closed separation region by free-stream turbulence. *J. Fluid Mech.* **446**, 271–308.

[39] KATZ, Y., NISHRI, B. & WYGNANSKI, I. 1989 The delay of turbulent boundary layer separation by oscillatory active control. *AIAA Paper 89–1027.*

[40] LIN, J. C. 1999 Application of micro-vortex generators for turbulent flow separation control. In *Mechanics of Passive and Active Flow Control* (ed. G. E. A. Meier & P. R. Viswanath), *Fluid Mechanics and its Applications*, vol. 53, pp. 81–88. Dordrecht: Kluwer Acad. Publ., proceedings of the IUTAM Symposium, 7.–11.9.98, Göttingen.

[41] MCCULLOUGH, G. B. & GAULT, D. E. 1951 Examples of three representative types of airfoil-section stall at low speeds. *Tech. Rep.* NACA TN 2502. Ames Aeronautical Laboratory, Moffett Field, Cal.

[42] MIAU, J. J., LEE, K. C., CHEN, C. R. & CHOU, J. H. 1991 Control of separated flow by a two-dimensional oscillating fence. *AIAA J.* **29**, 1140–1148.

[43] OBI, S., OHIZUMI, H., AOKI, K. & MASUDA, S. 1993 Turbulent separation control in a plane asymmetric diffuser by periodic perturbation. In *Eng. Turbulence Modelling and Expts.* (ed. W. Rodi & F. Martelli), , vol. 2, pp. 633–642. Elsevier Sci. Publ.

[44] PRANDTL, L. 1904 Über Flüssigkeitsbewegung bei sehr kleiner Reibung. In *Verh. III Int. Math. Kongr. Heidelberg*. Leipzig: Teubner Verlag, 1905.

[45] RECHENBERG, I. 1965 Skin-friction measurements in front of a two-dimensional obstacle by means of a surface fence. Private communication.

[46] RUDERICH, R. & FERNHOLZ, H.-H. 1986 An experimental investigation of a turbulent shear flow with separation, reverse flow, and reattachment. *J. Fluid Mech.* **163**, 283–322.

[47] SCHMIDT, J. J. 1997 Experimental and numerical investigation of separated flows. Ph. D. thesis, Danmarks Tekniske Universitet.

[48] Schubauer, G. B. & Spangenberg, W. G. 1960 Forced mixing in boundary layers. *J. Fluid Mech.* **8**, 10–32, and Nat. Bur. Stand. Report 6107 (1958).

[49] Seifert, A., Bachar, T., Koss, D., Shepshelovich, M. & Wygnanski, I. 1993 Oscillatory blowing: A tool to delay boundary-layer separation. *AIAA J.* **31** (11), 2052–2060.

[50] Seifert, A. & Wygnanski, I. 1994 The control of separation by periodic oscillations. *AIAA Paper 94-2608*.

[51] Sigurdson, L. W. 1995 The structure and control of a turbulent reattaching flow. *J. Fluid Mech.* **298**, 139–165.

[52] Sigurdson, L. W. & Roshko, A. 1988 The structure and control of a turbulent reattaching flow. In *Turbulence Management and Relaminarization* (ed. H. W. Liepmann & R. Narashima), pp. 497–514. Berlin: Springer-Verlag, IUTAM Symposium Bangalore, India, 1987.

[53] Siller, H. A. 1999 Reduction of the recirculation length downstream of a fence by an oscillating cross-flow. Dissertation, Technische Universität Berlin.

[54] Siller, H. A. & Fernholz, H.-H. 1997 Control of a separated flow downstream of a two-dimensional fence by low-frequency forcing. *Applied Sci. Res.* **57**, 309–318, Kluwer Acad. Publ.

[55] Siller, H. A. & Fernholz, H.-H. 1998 Projekt RaWid - Experimentelle Untersuchung einer ablösenahen turbulenten Grenzschicht mit aktiver Beeinflussung des Ablöseverhaltens. Techn. rep. Technische Informationsbibliothek Hannover, http://edok01.tib.uni-hannover.de/edoks/e001/302603123.pdf.

[56] Siller, H. A. & Fernholz, H.-H. 1999 Manipulation of separation on an airfoil by acoustic excitation. In *Mechanics of Passive and Active Flow Control* (ed. G. E. A. Meier & P. R. Viswanath), *Fluid Mechanics and its Applications*, vol. 53, pp. 317–322. Dordrecht: Kluwer Acad. Publ., proceedings of the IUTAM Symposium, 7.–11.9.98, Göttingen.

[57] Siller, H. A. & Fernholz, H.-H. 2000 Trailing edge separation control on an airfoil in low speed flow. *ZAMM* **80**, 89–92, suppl.

[58] Siller, H. A. & Fernholz, H.-H. 2001 Separation behaviour in front of a two-dimensional fence. *Eur. J. Mech. B/Fluids* **20**, 727–740.

[59] Smith, B. L. & Glezer, A. 1998 The formation and evolution of synthetic jets. *Phys. Fluids* **10**, 2281–2297.

[60] Taylor, H. D. 1948 Design criteria for and application of the vortex generator mixing principle. Rep. M-15038–1. United Aircraft Corp.

[61] Urzynicok, F. 1997 Separation control on an airfoil under post-stall conditions using low-frequency acoustic and low-frequency mechanical excitation. Diplomarbeit, Technische Universität Berlin, Hermann-Föttinger-Institut für Strömungsmechanik.

[62] Urzynicok, F. & Fernholz, H.-H. 1999 Separation control on an airfoil under post-stall conditions by mechanical excitation. In *Mechanics of Passive and Active Flow Control* (ed. G. E. A. Meier & P. R. Viswanath), *Fluid Mechanics and its Applications*, vol. 53, pp. 249–254. Dordrecht: Kluwer Academic Publishers, proceedings of the IUTAM Symposium, 7.–11.9.98, Göttingen.

[63] Urzynicok, F. & Fernholz, H.-H. 2002 Flow-induced acoustic resonators for separation control. *AIAA Paper 2002-2819*.

[64] Wengle, H., Huppertz, A., Bärwolff, G. & Janke, G. 2001 The manipulated transitional backward-facing step flow: an experimental and direct numerical simulation investigation. *Eur. J. Mech. B/Fluids* **20**, 25–46.

[65] WERLÉ, H. 1960 Essais de soufflage au tunnel hydrodynamique a visualisation. Note Technique 61. ONERA.

[66] WILLIAMS, D. R., ACHARYA, M., BERNHARDT, J. & YANG, P.-M. 1991 The mechanism of flow control on a cylinder with the unsteady bleed technique. *AIAA Paper 91–0039*.

[67] WU, J. M. & WU, J. Z. 1992 Vortex lift at a very high angle of attack with massively separated unsteady flow. In *IUTAM Symp. on Fluid Dynamics of High Angle of Attack*. Tokyo, Japan.

[68] WYGNANSKI, I. 1997 Boundary layer and flow control by periodic addition of momentum. *AIAA Paper 97–2117*.

[69] ZAMAN, K. M. B. Q., BAR-SEVER, A. & MANGALAM, S. M. 1987 Effect of acoustic excitation on the flow over a low-Reynolds-number airfoil. *J. Fluid Mech.* **182**, 127–148.

[70] ZHOU, M. D., CHENG, X., FERNHOLZ, H.-H. & DENGEL, P. 1990 Experimental research on active separation control for lift augmentation and drag reduction. Data rep. HFI Inst. Rep. 01/90. Technische Universität Berlin.

[71] ZHOU, M. D., FERNHOLZ, H.-H., MA, H. Y., WU, J. Z. & WU, J. M. 1993 Vortex capture by a two-dimensional airfoil with a small oscillating leading-edge flap. *AIAA Paper 93–3266*.

[72] ZHOU, M. D., GOETZ, M., KALTER, M. & FERNHOLZ, H.-H. 1999 The influence of periodoc forcing on a laminar airfoil with two slots. Data rep. Technische Universität Berlin.

Aerodynamic Flow Control Using Synthetic Jet Actuators

Michael Amitay[1] and Ari Glezer[2]

[1] Aerospace, Transportation and Advanced Systems Laboratory
[2] Woodruff School of Mechanical Engineering, Georgia Institute of Technology, Atlanta, GA 30332-0405

Abstract. The suppression of post-stall separation over an unconventional 2-D airfoil at moderate Reynolds numbers (up to 10^6) using synthetic (zero net mass flux) jet actuators is discussed. As shown by the authors in earlier investigations, the apparent modification of the surface shape by the interaction domain between the actuator jets and the cross flow results in a local displacement of the cross flow streamlines. The concomitant modification of the streamwise pressure gradient upstream of where the flow nominally separates in the baseline configuration can lead to complete suppression of separation over a significant range of angles of attack in the post-stall domain. While in the absence of flow control the airfoil is stalled at angles of attack exceeding 5°, actuation leads to either completely or partially attached flow within the entire range of angles tested (up to 25°) that is accompanied by a dramatic increase in lift and a corresponding decrease in pressure drag. Actuation is typically effected at frequencies that are *an order of magnitude higher* than the characteristic (shedding) frequency of the airfoil [i.e., $F^+ \sim O(10)$ rather than $F^+ \sim O(1)$]. When the actuation frequency F^+ is O(1), the reattachment is characterized by a Coanda-like tilting of the separated shear layer and the formation of large vortical structures at the driving frequency that persist beyond the trailing edge of the airfoil and lead to unsteady attachment and consequently to a time-periodic variation in vorticity flux and in circulation. In contrast, the suppression of separation at high actuation frequencies [i.e., $F^+ = O(10)$] is marked by the absence of organized vortical structures along the flow surface. The dynamics of the transient lift in controlled reattachment and separation are investigated using pulsed amplitude modulation of the actuation input and is exploited to improve the efficacy of the jet actuators by using pulse modulation of the excitation input.

1 Introduction

Optimum aerodynamic performance of aircraft wings that has been traditionally achieved by appropriate design of the airfoil section to avoid flow separation may be somewhat compromised when the wing design is driven by non-aerodynamic constrains (e.g., payload, or stealth). Hence, these designs may rely on either active or passive flow control techniques to prevent flow separation and to maintain acceptable aerodynamic performance throughout the normal flight envelope. Although the relative simplicity of passive flow control devices (e.g. vortex generators, oscillating cavities, etc.) makes them attractive means for delaying flow separation, they afford no proportional control and introduce a drag penalty in the absence of "natural" separation. In contrast, conventional active control approaches which are based on the manipulation of fundamental instabilities of the separated free shear layer and derive their effectiveness from its receptivity to relatively low-level actuation input can be largely innocuous except when activated and have the potential for delivering variable power.

Controlled reattachment of separated flows over lifting surfaces at moderate and high angles of attack with the objective of improving aerodynamic performance and extending the flight envelope has been the focus of a number of investigations since the early eighties. Active control techniques that have achieved varying degrees of separation control by manipulation of the unstable separated free shear layer have included external and internal acoustic excitation (e.g., Ahuja and Burrin [1] and Zaman et al. [2]), vibrating ribbons or flaps (e.g., Bar-Sever [3]), and steady and unsteady blowing/bleed (e.g., Seifert et al. [4] and Williams [5]).

The reattachment of a separated shear layer on an stalled airfoil by means of internally-driven acoustic excitation of the boundary layer upstream of separation (through a small rectangular orifice near the leading edge) was first investigated by Huang et al. [6] and Hsiao et al. [7, 8] (on NACA 63_3-018 airfoil). While the actuation frequency used by Huang et al. [6] was limited to the shedding frequency of the airfoil (i.e., dimensionless frequency $F^+ \sim 1$, scaled with the separated flow region), Hsiao et al. [7] reported similar lift recovery over a much broader range of excitation frequencies (up to $F^+ = 20$) that *far exceed the unstable frequency of the separating shear layer*. In the follow-on work of Chang et al. [9], the actuation level was quantified using the orifice momentum coefficient (based on the amplitude of the velocity oscillations) and the authors reported that actuation at $F^+ = 2$ and $C_\mu < 10^{-4}$ led to a post-stall increase in lift of up to a 50%. However, as noted in the earlier work of Hsiao et al. [7] separation could also be controlled up to $F^+ < 20$ at higher actuation levels.

Flow control work in the 50s showed that separation on lifting surfaces could be avoided or at least delayed by the streamwise injection of fluid near the surface using a tangential wall jet. The work of Seifert et al. [4] demonstrated that separation over a flap (at 25% chord of a NACA 0015 airfoil) can be effectively controlled with substantially less mass flux by replacing the continuous wall jet at the flap hinge with oscillatory injection at the unstable frequency of the separating shear layer. These authors reported that oscillatory blowing at reduced frequencies within the range $1 < F^+ < 3$ combined with a low-level steady blowing yielded effective actuation for maximum aerodynamic performance augmentation. Similar approach was used in a later investigation (Seifert et al. [10]) to control leading edge separation on the same airfoil where the dimensionless frequency for correlating C_μ was scaled with the length over which reattached flow was maintained. In a related munerical simulation of the post-stall ($\alpha = 22°$) flow about an NACA0012 airfoil actuated at $F^+ \cong 1$ by a zero net mass flux jet, Donovan et al. [11] demonstrated that the increased lift had substantial oscillations (up to 20% peak-to-peak) at the actuation frequency.

The work described here is part of an ongoing investigation of the flow mechanisms of a novel control approach of the performance of lifting surfaces through fluidic modification of their apparent aerodynamic shape. Control is effected by surface-mounted fluidic actuators based on synthetic jet technology having the distinct feature that they do not require a fluid source and are formed locally from the fluid adjacent to the flow surface in which they are embedded (Smith and Glezer [12]). Thus, the interaction between these jets and the cross flow over the surface leads to the formation of a domain that displaces the local streamlines of the cross flow thereby inducing an 'apparent' or 'virtual' change in the shape of the surface and an *apparent* modification of the flow boundary and the local streamwise pressure gradient.

While conventional flow control schemes have primarily focused on the mitigation of flow separation by coupling to the instability of the separating shear layer on the time scale of the flow about the airfoil, fluidic modification of the apparent aerodynamic shape of aero-surfaces does not necessarily rely on coupling to global flow instability and thus can be applied over a broader range of flow conditions. Furthermore, this approach to flow control can accommodate broader band control algorithms because the actuation frequency is typically at least an order of magnitude higher than the characteristic frequency of the flow. Smith et al. [13] and Amitay et al. [14, 15] demonstrated the utility of synthetic (zero mass flux) jet actuators operated at $F^+ = O(10)$ for the suppression of separation over an unconventional airfoil at moderate Reynolds numbers (up to 10 [6]) and consequently a substantial increase in lift and decrease in pressure drag. Recent experiments by Erk [16] demonstrated suppression of separation on an FX61-184 airfoil at Reynolds numbers up to 3×10^6 using synthetic jet actuation at frequencies up to $F^+ \sim O(100)$. As shown by Amitay et al. [17], the quasi-steady lift coefficient of the attached flow can be further augmented by exploiting the flow transients associated with the onset of separation using a prescribed unsteadiness of the locally separated flow domain that is induced by a temporally-modulated control input.

The present paper reviews some of the elements of the work on the modification of the aerodynamic performance of the unconventional airfoil of Amitay et al. [14, 15, 18] using fluidic control that is effected by synthetic jet actuators near the airfoil's leading edge (Sect. 3). The control effectiveness at actuation frequencies that are either on of the order of or well above the natural shedding frequency of the airfoil [i.e., $F^+ \sim O(1)$ and $O(10)$] are discussed in Sect. 4. The mechanisms of flow transients that are associated with the reattachment and separation processes are discussed in Sect. 5. Finally, Sect. 6 describes how these reattachment and separation flow transients can be exploited to augment the actuation efficiency by using pulsed modulated actuation input.

2 Experimental Apparatus and Procedure

The application of synthetic jets for separation control over an unconventional airfoil are demonstrated using a thick airfoil (Fig. 1) model that was developed in collaboration with Boeing's Phantom Work Group in St. Louis (Smith et al. [13]). The airfoil's leading edge is formed by a 62 mm circular cylinder mounted within an aerodynamic fairing that is based on a uniformly stretched NACA four-digit series symmetric airfoil for which the thickness to chord ratio is $t/c = 0.24$ ($c = 254\,\mathrm{mm}$).

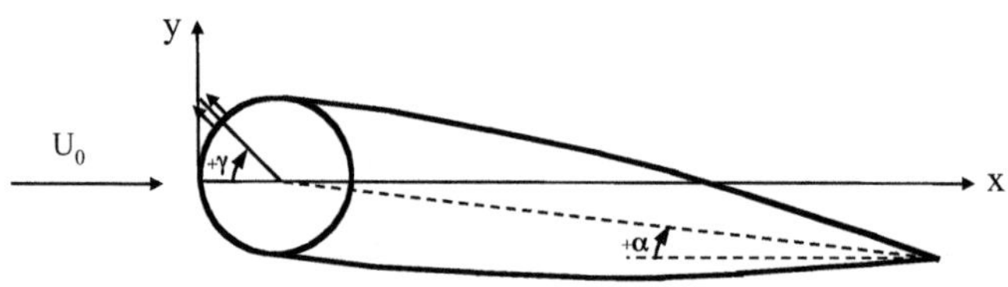

Fig. 1. Unconventional airfoil model

The fairing section is laminated using a rigid foam material with an epoxy-fiberglass overlay. The surfaces of the cylinder and of the fairing are well polished and the transition between the surface of the cylinder and the edge of the fairing is designed to have a close fit such that the surface discontinuity at the edge of the overlap domain is less than 0.1 mm (and is too small to trigger local separation). The gap between the cylinder and fairing is specifically designed to prevent leakage between the pressure and suction sides of the airfoil, and at a given angle of attack and actuator jet angle, the junction between the cylinder and the fairing is sealed with thin (less than 0.05 mm) tape.

The airfoil is mounted in the test section of an open-return, low-speed wind tunnel having a square test section measuring 91 cm on a side (the maximum air speed is 32 m/s with a free-stream turbulence level less than 0.25%). The upper and lower walls of the wind tunnel are adjusted to compensate for blockage created by the airfoil (wall interference effects are estimated to result in a 1.5–2% over-prediction of the drag coefficient). Distributions of surface pressure are measured using 47 pressure taps that are located in the spanwise mid-plane and are equally spaced circumferentially around the cylinder and additional 45 pressure taps at mid span along the top and bottom surfaces of the fairing.

The center section of the leading edge cylinder houses a pair of parallel synthetic jet actuators each having a flush-mounted 140 mm × 0.5 mm orifice that are 2.5 mm apart azimuthally and co-linear with the axis of the cylinder. The jet actuators can be positioned at an azimuthal angles γ between $-90°$ to $+90°$ (at $\alpha = 0°$) relative to the incoming flow direction by rotating the cylinder independently of the fairing. Two fences are placed at the edges of the actuator orifice to maintain a nominally two-dimensional flow in the controlled section of the airfoil. High-frequency synthetic jets are produced at $F^+ = 10, 14.7$ and 20 (f_a = 740 Hz, 1088 Hz and 1480 Hz, respectively) using piezo-electrically driven diaphragms mounted in compact shallow cavities underneath the surface of the cylinder. Low-frequency synthetic jets are produced through the same orifices (in the absence of the piezoelectric drivers) at $F^+ = 0.95, 2.05$, and 3.3 (f_a =71 Hz, 148 Hz, and 246 Hz, respectively) by conventional pressure speakers mounted on opposite ends of the cylinder cavity outside of the wind tunnel's test section.

In the present work, the performance of the actuators is measured using the momentum coefficient,

$$C_\mu = \frac{\bar{I}_j}{\frac{1}{2}\rho_o U_o^2 c}$$

where $\bar{I}_j$ is the time-averaged momentum flux per unit length during the outstroke and is given by

$$\bar{I}_j = \frac{1}{\tau}\rho_j b \int_0^{\tau} \langle u_j^2(\phi)\rangle d\phi \ ,$$

$\tau = T/2$ (T is the period of the diaphragm motion), ρ_j and ρ_o are the jet and free-stream fluid densities, respectively, b is the jet orifice width, c is the chord, U_o is the free-stream velocity, and $\langle u_j(\phi)\rangle$ is the phase-averaged velocity at the jet exit plane. The velocity within the jet orifice is measured using a miniature hot-wire sensor and the velocity traces are rectified when the velocity reverses its direction at mid-cycle. Thus, during the suction part of the cycle the velocity is inverted to reflect the correct flow direction following the procedure of Smith and Glezer [12].

Cross stream distributions of the streamwise and cross-stream velocity components are measured in the wake of the airfoil using X-configuration hot wire miniature sensors that are mounted on a computer-controlled traversing mechanism. The velocity and vorticity fields in the cross-stream (x-y) plane, $z = 0$, above the airfoil (i.e., on the suction side) are measured using Particle Image Velocimetry (PIV). The flow is seeded using smoke particles and it is illuminated using a double-pulse ND-YAG laser. Image pairs are captured using a 1008 × 1016 pixel CCD camera with a magnification of 53 µm/pixel (the nominal particle diameter is sub-pixel). Velocity vectors are computed on a 62 × 62 grid using a standard cross-correlation technique. All data are averaged over 300 realizations (150 image pairs).

3 Modification of the Time-Averaged Aerodynamic Performance

This section reviews briefly some of the effects of jet actuation on the aerodynamic performance (as measured by pressure distributions, lift and pressure drag) of the unconventional airfoil described in Sect. 2. The distributions of the static pressure coefficient about the circumference of the airfoil in the presence and absence of actuation (closed and open symbols) are shown in Figs. 2a-f ($\alpha = 0°$, $5°$, $10°$, $15°$, $20°$ and $25°$, respectively) for $\mathrm{Re}_c = 310,000$. The jet angle for the controlled flow is $\gamma = 60°$ (relative to the oncoming free stream), the jet momentum coefficient is $C_\mu = 3.5 \cdot 10^{-3}$, and $\hat{x} = x/c$.

In the absence of flow control, the airfoil stalls at angles of attack exceeding 5° (Fig. 2b). At $\alpha = 0°$ (Fig. 2a), the baseline flow is attached over the entire airfoil surface. These data show that while the suction peak is almost the same on both sides of the airfoil, there are slight differences in the domains where the fairing blends into the cylinder. Although for the present momentum coefficient the effect of the jets on the attached flow is small, it appears that despite the (top-to-bottom) asymmetry of the actuation the pressure distributions on the top (where the jets are located) and bottom surfaces become more similar when the jet are activated. For $\alpha \leq 15°$ (Figs. 2b-d), the flow becomes fully reattached when control is applied, and the resulting pressure distributions exhibit a large suction peak in the vicinity of the maximum thickness (near $\hat{x} = 0.1$). A rapid pressure recovery occurs for $0.1 < \hat{x} < 0.2$ followed by a more gradual pressure recovery towards the trailing edge.

It is interesting to note that at moderate angles of attack the pressure within some streamwise domains in the absence of actuation is higher on the upper (suction) surface than on the lower (pressure) surface of the airfoil resulting in a reduction in lift (e.g., $0.075 < \hat{x} < 0.30$ and $0.06 < \hat{x} < 0.22$ in Figs. 2b and 2c, respectively). However, when the jets are activated and the flow becomes fully attached, the pressure on the upper surface is continuously lower than on the lower surface from the leading edge and along most of the airfoil (through $\hat{x} = 0.62$ for $\alpha = 5°$ and $\hat{x} = 0.72$ for $\alpha = 10°$). The existence of small domains near the trailing edge in which the pressure difference between the two surfaces is reversed indicates that the streamwise rate of pressure recovery resulting from the reattached flow on the upper surface exceeds the rate of recovery associated with the curvature of the bottom surface. This effect decreases with increasing angle of attack where the pressure distribution on the lower surface becomes almost invariant with $\hat{x}$ (Figs. 2d and 2e) and finally decreases with $\hat{x}$ (Fig. 2f).

At $\alpha = 20°$ (Fig. 2e), the jet actuation produces only partial reattachment. The flow remains attached through $\hat{x} = 0.22$ but subsequently separates and the pressure distribution within the separated flow domain is almost identical to the corresponding distribution in the absence of control. When $\alpha = 25°$ (Fig. 2f), the baseline flow separates farther upstream ($\hat{x} = 0.02$) and as a result the pressure within the separated flow domain is actually higher than

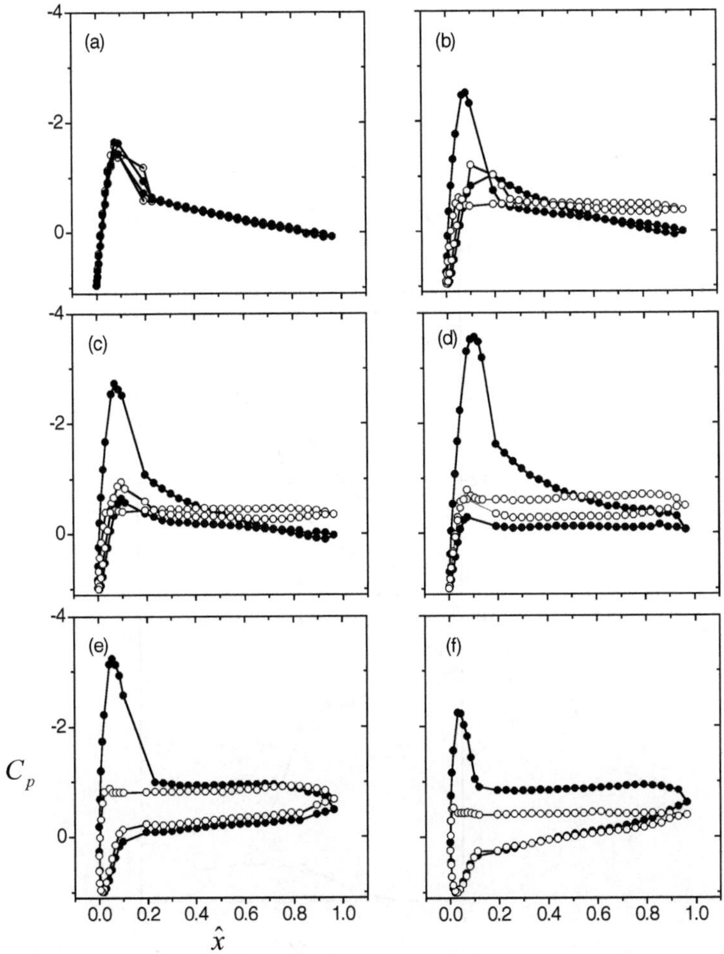

Fig. 2. Pressure coefficient distributions around the airfoil at different angles of attack. Forced (—•—), and unforced (—o—). $\alpha = 0°$ (**a**), $\alpha = 5°$ (**b**), $\alpha = 10°$ (**c**), $\alpha = 15°$ (**d**), $\alpha = 20°$ (**e**) and $\alpha = 25°$ (**f**)

for $\alpha = 20°$. However, even though the attained suction peak in the presence of actuation is lower than at $\alpha = 20°$, separation still occurs at the same streamwise location ($\hat{x} = 0.2$) and as a result the pressure in the separated region of the actuated flow is considerably lower than in the baseline flow. While the increase in the pressure difference (compared to the baseline flow) results in increased lift, at these angles of attack there is also a significant increase in pressure drag. Nevertheless, as shown in Fig. 3 below, the lift-to-pressure drag ratio increases compared to the baseline case.

The pressure distribution on the airfoil is integrated to yield the lift coefficient, the pressure drag coefficient and the lift-to-pressure drag ratio over

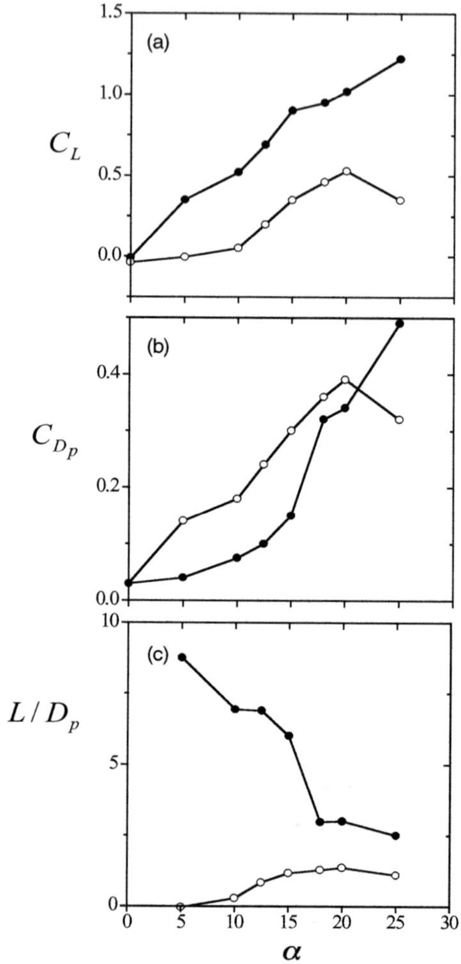

Fig. 3. Lift coefficient (**a**), pressure drag coefficient (**b**) and lift-to-pressure drag ratio (**c**) vs. angle of attack, α for $\gamma = 60°$. Forced (—●—) and unforced (—o—)

a range of angles of attack (Figs. 3a-c, respectively). The corresponding distributions of the baseline flow are also shown for comparison (open symbols). Without control, the aerodynamic performance of this airfoil is rather poor (e.g., the lift-to-pressure drag ratio at $\alpha = 5°$ is 0.115), and the airfoil is stalled even at low angles of attack. However, with actuation ($\gamma = 60°$, $C_\mu = 3.5 \cdot 10^{-3}$), the lift coefficient increases almost linearly with angle of attack as might be expected for a conventional airfoil. Furthermore, the attached flow in the presence of actuation results in a substantial reduction in the pressure drag coefficient. For example at $\alpha = 5°$, the pressure drag decreases by 45% with respect to the baseline case. Even at relatively high angles of attack ($\alpha = 25°$), where the flow is only partially attached, the lift curve has yet to exhibit the downturn characteristic of stall. However, the monotonically increasing pressure drag (which exceeds the drag coefficient of the baseline airfoil at $\alpha = 25°$) and the pressure distribution (Fig. 2f) suggest that stall is imminent. Figure 3c shows that synthetic jet actuation significantly improves the lift-to-pressure drag ratio through the entire range of angles of attack in the present investigation. At $\alpha = 5°$, this ratio is increased by almost *two orders of magnitude*. However, the improvement in the L/D_p diminishes as the angle of attack increases.

4 The Role of the Actuation Frequency

As noted in Sect. 1, much of the earlier work that focused on extension of the pre-stall envelope of various conventional airfoils by utilized active (time-periodic) actuation upstream of separation. The characteristic period of the actuation typically has been comparable to the time of flight over the separated flow domain corresponding to a dimensionless frequency F^+, which is the ratio between the latter and former time scales, of order 1 (i.e., the actuation frequency is comparable to the shedding frequency of the airfoil). The natural receptivity of the separating shear layer to this frequency band results in a Coanda-like tilting of the shear layer towards the surface of the airfoil and therefore in partial restoration of the lift. In contrast to this approach, in the present work the actuation frequency is selected to be at least an order of magnitude higher than the "natural" (or shedding) frequency of the airfoil in order to bypass flow separation altogether. Furthermore, as shown in Sects. 5 and 6, the broadening of the actuation bandwidth enables coupling to transient effects within the flow that can be exploited for enhancement of the actuation effectiveness (e.g., by reducing the required C_μ).

To demonstrate the effect of the actuation frequency on the control effectiveness, the synthetic jet actuators are activated at six frequencies corresponding to $F^+ = 0.95$, 2.05, 3.4, 10, 14.7 and 20 (71 Hz, 148 Hz, 246 Hz, 740 Hz, 1088 Hz and 1480 Hz, respectively). At $F^+ = 0.95$, 2.05 and 3.4, the jets are driven by audio speakers while at high frequencies the jets are formed by compact, piezo-electrically-driven actuators (cf. Sect. 2). The actuators are calibrated at each operating frequency outside of the wind tunnel, and then once again in the presence of a cross flow in the wind tunnel. It is found that when the operating frequencies are below $F^+ < 4$, the jets driven by the speaker actuators are spanwise-uniform (along the long dimension of the orifice). However, at higher frequencies the speaker-driven jets develop spanwise non-uniformities due to three-dimensional acoustic effects within the actuator cavity. Velocity measurements (not shown here) within the actuator orifice using a miniature hot-wire sensor reveal that for $F^+ > 4$, the performance of the speaker-driven actuator jets is markedly different in the presence and absence of a cross flow. In fact, in the presence of a cross flow, the velocity of the speaker driven jets at a given spanwise position along the orifice decreases by more than 80% and thus their ability to affect the separating shear layer is substantially reduced. In contrast, the velocity of the high-frequency jets produced by the piezoelectric actuators is spanwise-uniform, virtually invariant with tunnel speed and leads to a complete reattachment of the separated flow.

Distributions of the pressure coefficient around the airfoil for $F^+ = 0.95$, 2.05 and 3.4 and $F^+ = 10$, 14.7 and 20 are shown in Fig. 4a and Fig. 4b, respectively (the shedding frequency of the baseline flow is $F^+ = 0.7$). Here, $Re_c = 310,000, \gamma = 60°$, $C_\mu = 3.5 \cdot 10^{-3}$ and $\alpha = 17.5°$. The pressure distribution for the stalled flow (solid line) is also shown for reference. Actuation at low frequencies (Fig. 4a) results in a very sharp suction peak near $\hat{x} = 0.075$, which corresponds to the location of the separation in the baseline flow. Concomitantly, the degree of pressure recovery towards the trailing edge is reduced with increasing control frequency, leading to an increase in pressure-drag. The pressure distributions for $F^+ \geq 10$ (Fig. 4b), which is over an order of magnitude higher than the shedding frequency ($F^+ = 0.7$), are significantly different and exhibit larger and wider suction peaks with a corresponding larger increase in the lift coefficient. Following the suction peak the pressure difference between the suction and pressure sides is smaller than at corresponding stations at the low actuation frequencies resulting in lower pressure drag. Moreover, while at low actuation frequencies (Fig. 4a) the pressure distribution varies with F^+, at high actuation frequencies (Fig. 4b) the pressure distribution appears to be independent of F^+, suggesting that when the actuation frequency is high enough, the details of the flow reattachment become independent of the frequency.

The variation of the ratio of lift to (pressure) drag, L/D_p, with actuation frequency is shown in Fig. 5. Two distinct domains are immediately apparent. In the first domain (I), where the actuation frequencies are of the same order

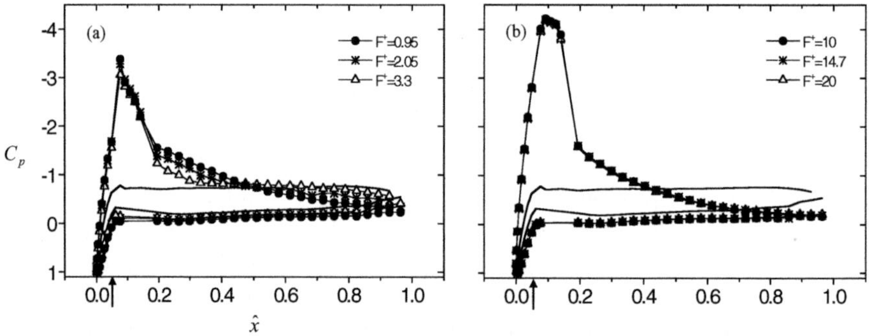

Fig. 4. Variation of the pressure coefficient with the dimensionless forcing frequency; $F^+ \sim O(1)$ (**a**) and $F^+ \sim O(10)$ (**b**). (The distribution for the unforced flow is shown in a solid line)

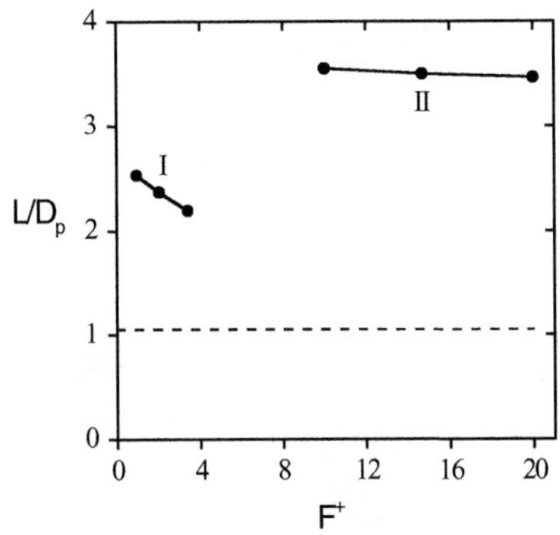

Fig. 5. The variation of the lift-to-pressure drag ratio with dimensionless actuation frequency, F^+ for $\gamma = 60°$. The unforced value is shown in the dashed line

as the shedding frequency ($F^+ < 4$), L/D_p decreases with increasing F^+ (from 2.65 for $F^+ = 0.95$ to 2.35 for $F^+ = 3.3$), which may be the result of reduced receptivity of the separated shear layer. In the second domain (marked as II on the figure), the actuation frequency is at least an order of magnitude higher than the shedding frequency (i.e., $F^+ \geq 10$) and the lift to pressure drag ratio is higher ($L/D_p \approx 3.2$). It appears that L/D_p is almost independent of the actuation frequency, suggesting that the mechanism that leads to the suppression of separation is not associated with the stability of the separated shear layer. Note that bandwidth limitations of the low- and high-frequency actuators prevented overlap within the domain $4 < F^+ < 10$.

The nominally two-dimensional flow field associated with each of the two frequency bands is computed from a sequence of PIV images that are captured in the x-y plane ($z = 0$) above the suction side of the airfoil. Each PIV data set is comprised of three partially overlapping frames measuring 100 mm on the side where time- or phase-averaged velocity distributions are computed from an ensemble of 150 image pairs.

Figures 6a-g show cross-stream maps of the velocity vector distributions. The figure includes a map of the time-averaged baseline flow (Figs. 6a) followed by pairs of phase- and time-averaged maps of the actuated flow for $F^+ = 0.95$, 3.3 and 10. Each pair includes a phase-averaged velocity field (Figs. 6b, 6d and 6f) in which the data is taken phase-locked to the actuation waveform, and a time-averaged field (Figs. 6c, 6e and 6g). In the absence of control, the velocity vector map of the time-averaged baseline flow (Fig. 6a) exhibits a large recirculating flow domain with reversed flow above the surface of the airfoil. Figure 6b ($F^+ = 0.95$) clearly shows that the separating shear layer is effectively tilted towards the surface resulting in the rollup and advection of coherent vortical structures at the actuation frequency (having a characteristic wavelength of $\sim 0.4c$). In fact, the eddy that precedes the

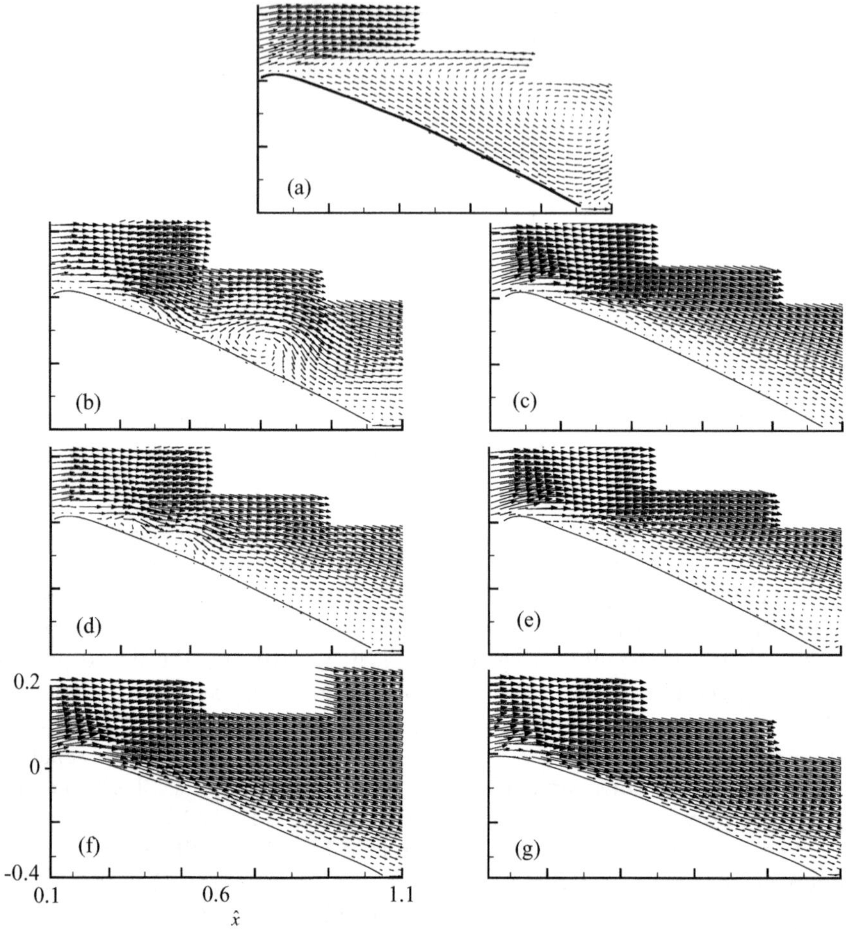

Fig. 6. Cross-stream maps of velocity vectors. unforced (**a**), phase- and time-averaged map pairs of the forced flow: $F^+ = 0.95$ (**b**, **c**), 3.3 (**d**, **e**) and 10 (**f**, **g**)

vortex centered at $\hat{x} = 0.7$ is already within the wake of the airfoil. The time-averaged flow (Fig. 6c) exhibits a broad region of low-velocity near the surface, and the time-averaged flow is actually *reversed* near the surface of the airfoil, indicating that the flow is not fully attached there. When the excitation frequency is increased to $F^+ = 3.3$ (Figs. 6d and e), the streamwise wavelength of the shear layer vortices decreases to approximately 0.15c and the coherent structures appear to lose their phase coherence (relative to the actuation waveform) around $\hat{x} = 0.7$, although they are clearly present in the individual snapshots. Again, the velocity vector maps of the time-averaged flow suggest that the flow is not fully attached and there is a region of reversed flow near the surface that is induced by the passage of the large vortical structures.

For actuation frequency that is an order of magnitude higher than the "natural" baseline frequency of the separating shear layer (Figs. 6f and 6g), the phase coherence of the structures within the attached flow is greatly reduced compared to the lower actuation frequencies, and the boundary layer of the actuated flow appears to be fully attached over the entire length of the airfoil. In fact, the time- and phase-average vector fields are almost identical. Comparison of the time-averaged velocity profiles for the actuated flow (Figs. 6c, e and g) suggest that at low actuation frequencies the boundary layer of the mean flow is reminiscent of a thick separating boundary layer in an adverse pressure gradient while for $F^+ \sim 10$ the boundary layer is significantly thinner. The data in Figs. 6b and d also indicate that the time-periodic formation and shedding of the vortical structures can lead to synchronous time-periodic variation in the circulation and consequently in the lift (and drag) forces as shown in Fig. 15 below and is also confirmed in the numerical simulations of Donovan et al. [11] and Wu et al. [19].

The power spectra of the streamwise velocity measured at $\hat{x} = 2$ on the lower side of the airfoil's wake at a cross-stream elevation where the streamwise velocity deficit is half the maximum deficit are shown in Figs. 7a-d (the spectrum of the baseline flow is reproduced, using a dashed line, in each of the figures for reference), where the dimensionless frequency is $v^+ = (f \cdot c)/U_0$. As evident from the cross-stream distributions of the velocity vector maps (Fig. 6) when the flow is actuated (and nominally attached to the upper surface of the airfoil) the cross-stream extent of the wake is reduced substantially and is accompanied by a reduction in the magnitudes of rms velocity fluctuations within the wake. At $F^+ = 0.95$ (Fig. 7a) there is a strong spectral component at the actuation frequency and the entire spectrum appears to be attenuated by approximately 3.5. As F^+ is increased to 2 and 3.3, the spectral peak at the actuation frequency shifts towards the decaying part of the spectrum (it is virtually indistinguishable from the background at $F^+ = 3.3$) and the attenuation imposed by the collapse of the separated flow domain increases to 4.5 and 7, respectively. The spectrum of the actuated flow at $F^+ = 10$ (Fig. 7d) is remarkably different from the spectral distributions at the lower actuation frequencies. There is a stronger attenuation (well over an order of magnitude) at both the low and high spectral ends, and the spectrum includes a distinct inertial sub-range for $\nu^+ > 3$ over almost two decades (which includes the actuation frequency). The reduced power at all spectral components (compared to actuation at lower F^+) indicates that the attached flow removes less energy from the uniform stream resulting in a lower drag.

The effect of the actuation frequency on the flow structure over the present airfoil was also investigated numerically by Palaniswamy [20] using a novel limited numerical scales (LNS) approach, which is effectively a hybrid between RANS and LES approaches. In LNS the small-scale flow features are resolved wherever computational grid is adequate, and diffused in coarse parts of the

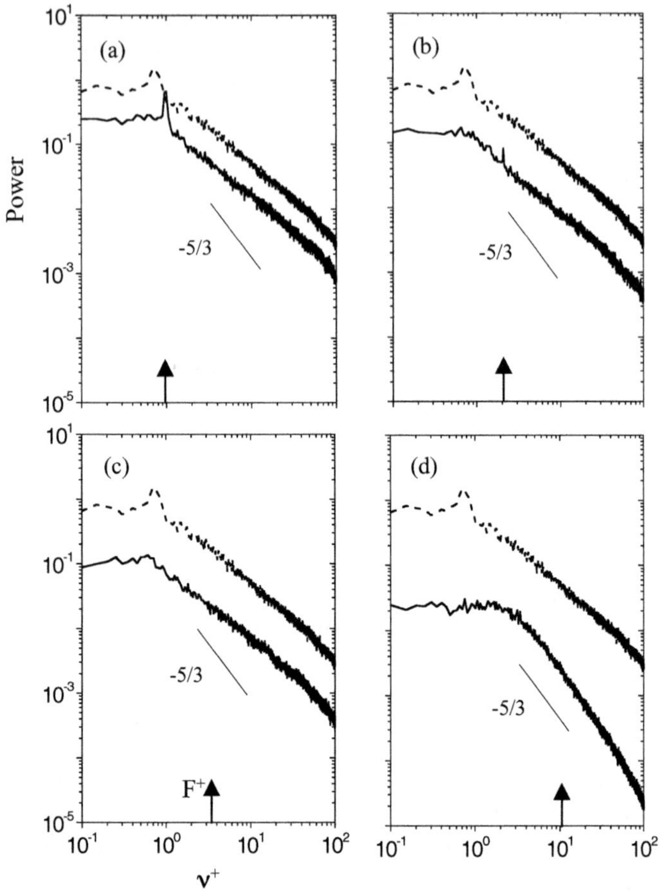

Fig. 7. Power spectra for $F^+ = 0.95$ (**a**), 2.05 (**b**), 3.3 (**c**), and 10 (**d**). The spectrum of the unforced flow is repeated in dashed line in each frame

mesh. The computations presented here are based on a CFD^{++} flow solver with a hybrid RANS/LES turbulence closure. Figures 8a and b are each a superposition of a map of the flow streamlines and a color raster plot of the pressure coefficient ($\mathrm{Re}_c = 310,000, \alpha = 17.5°$ and $\gamma = 60°$) for the baseline and actuated ($F^+ = 10$ and $C_\mu = 3.5 \times 10^{-3}$) flows, respectively. As shown in Fig. 8a, in the absence of actuation the airfoil is completely stalled and the image also shows the formation of a clockwise (CW) vortex near the leading edge and the previous CW vortex within the wake. Note also the formation of a counter clockwise vortex (CCW) near the trailing edge of the airfoil. It is remarkable that when the jets are activated (Fig. 8b), the flow becomes completely attached with a narrow wake and there is no discernible shedding of large-scale vortical structures. The attached flow is also characterized by the appearance of a low pressure domain (corresponding

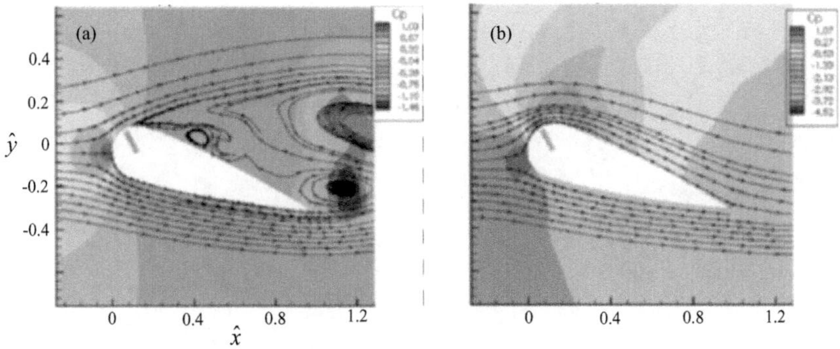

Fig. 8. Superposition of the flow streamlines and a color raster plot of the pressure coefficient. Baseline (**a**) and actuation at $F^+ = 10$ (**b**)

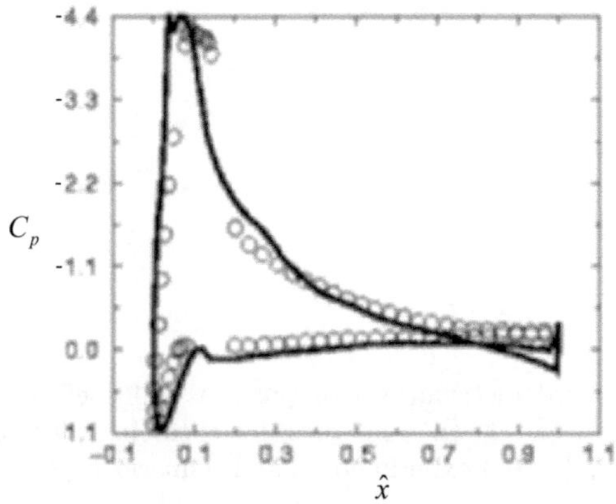

Fig. 9. Pressure coefficient distribution around the airfoil. $F^+ = 10$. (—o—) experimental, and (——) numerical

to the suction peak in Fig. 4b) on the upper surface of the airfoil near the leading edge. The LNS results are validated by comparing the numerical and measured pressure distributions around the airfoil (Fig. 9). The agreement between the results emphasizes the importance of resolving and capturing the spatial and temporal variations of the small-scale motions that are an important feature of LNS for simulating the effect of small-scale actuation devices on the macro scale flow structure.

An important consequence of the LNS simulations is the ability to capture the instantaneous pressure distribution on the surface of the airfoil during the actuation cycle at relatively low actuation frequencies (e.g., $F^+ < 4$). Figures 10a, b are phase-averaged streamlines maps and color raster plots of the on the pressure coefficient at two instances during the actuation cycle for $F^+ = 3.3$. These images clearly show the advection of a CW vortex (which is formed time-periodically at the actuation frequency). As noted in Sect. 1 and

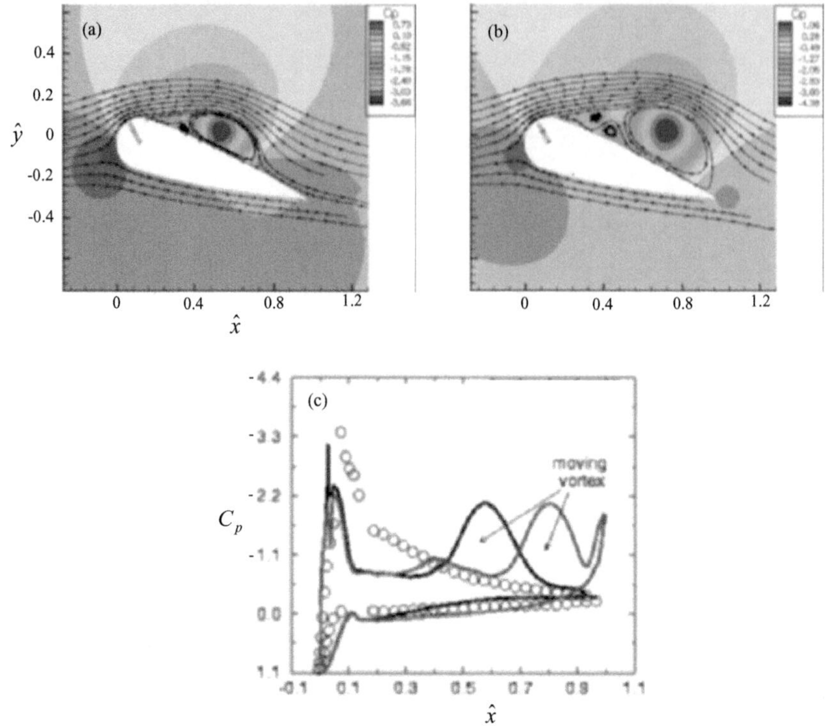

Fig. 10. Superimposed distributions of the pressure coefficient and streamlines around the airfoil, t_1 (**a**), t_2 (**b**) and pressure coefficient distribution around the airfoil (**c**). $F^+ = 3.3$. (—o—) experimental, and numerical: (——) t_1 and (——) t_2

in earlier numerical investigators (e.g., Donovan et al. [11] and Wu et al. [19]), the recovery of lift at low F^+ results in the tilting of the separating shear layer towards the surface of the airfoil and by the advection of large-scale vortical structures along the surface that may result in unsteady aerodynamic forces. The instantaneous pressure distributions about the surface of the airfoil corresponding to the data in Figs. 10a and b are shown in Fig. 10c along with the measured time-averaged pressure distribution (for $F^+ = 3.3$). The surface pressure distributions show domains of low pressure (suction) that are comparable to the instantaneous suction peaks near the leading edge of the airfoil and are somewhat weaker than the measured time-averaged peak. The time-periodic advection of these suction peaks contribute to the momentary lift and drag forces on the airfoil and clearly contribute to oscillations in the pitching moment (as discussed further in Sect. 5 below).

5 Elements of the Dynamics of Flow Reattachment and Separation

The dynamic response of controlled reattachment and separation at a post-stall angle of attack ($\alpha = 17.5°$) is investigated using pulsed amplitude modulation of the actuator (control) input at actuation frequencies that are of the order of and an order of magnitude higher than the natural frequency of the separating shear layer. The modulation is synchronized with the actuator's driving signal such that the leading edge of the modulating waveform coincides with a zero crossing of the actuator signal and continues for 0.5 sec (370 T). The flow transients resulting from this pulsed excitation are measured in detail across the near wake ($\hat{x} = 2$) using x-wire anemometry (in what follows, $\gamma = 60°$, $C_\mu = 3.5 \cdot 10^{-3}$ and $Re_c = 310,000$).

The transient effects of pulsed modulation are captured in color raster plots of the phase-averaged cross-stream distributions of the dimensionless spanwise vorticity $\langle\hat{\Omega}\rangle$ that is computed from the streamwise and cross-stream velocity components (Fig. 11). The jet actuators are driven at $F^+ = 10$ and the flow is unforced before and after the modulation is applied (marked with "up" and "down" arrows on the time scale). The flow transients associated with the onset of the modulation are sensed at the measurement station at $\hat{t} = 100$ (approximately 25 actuator periods after the modulation is effected). When the flow is separated (i.e., before and after the pulse modulated excitation is applied), the vorticity distribution in the wake is comprised of a train of vortical structures of alternating sign (clockwise vorticity is taken to be neg-

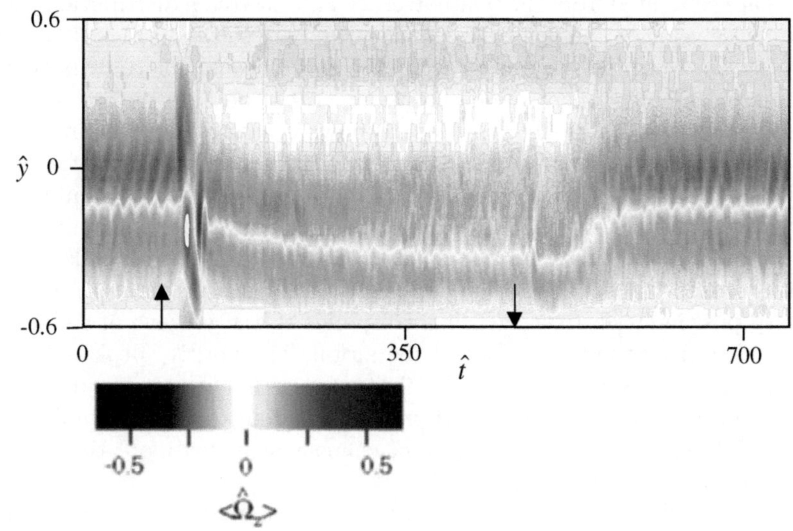

Fig. 11. Phase-averaged color raster plot of the cross-stream distribution of the spanwise vorticity; $F^+ = 10$

ative) having a nominal passage frequency of 50 Hz. Nevertheless, the total vorticity flux across the wake during one period of the ('natural') shedding frequency is approximately zero. As shown in Sect. 3, the actuation leads to flow reattachment and the establishment of a higher (positive) lift force on the airfoil, which must be accompanied by a change in the vorticity flux and a net increase in circulation associated with positive (counter-clockwise) vorticity. However, following the reattachment, a strong clockwise vortex indicating *a reduction in lift* is initially advected past the measurement station followed closely by a stronger counter-clockwise vortex indicating the re-establishment of lift. These two large vortices are followed by a series of smaller vortices of alternating signs and diminishing strength. It appears that the reduced wake of the attached airfoil ultimately reaches a state of symmetric vorticity distribution as can be seen for $\hat{t} > 300$ in Fig. 11. When the pulse modulation control is turned off, the flow separates again and the airfoil loses its lift. This reduction in lift is accompanied by a decrease in circulation and the shedding of negative (clockwise) vorticity. However, immediately following the termination of the control, a counter-clockwise vortex indicating *a momentary increase in lift* is advected past the measurement station before the separated vorticity field is established.

As noted by Amitay et al. [15], the time rate of change of the circulation is given by the phase-averaged vorticity flux

$$\frac{d\langle\hat{\Gamma}\rangle}{d\hat{t}} = \int_{\infty}^{\infty} \langle\hat{U}\rangle \cdot \langle\hat{\Omega}_{\mathrm{z}}\rangle d\hat{y}$$

and the incremental change in the circulation with respect to the baseline case, $\Delta\langle\hat{\Gamma}\rangle$, are estimated from the phase-averaged cross-stream distributions of the streamwise velocity and spanwise vorticity, and are shown in Figs. 12a and b, respectively. Note that the integration does not account for contributions of the fluctuating components, and it is assumed that because the measurement station is located only one chord-length downstream of the trailing edge of the airfoil, which is shorter than the wavelength of the shedding frequency, the interaction between successive vortices within this domain is minimal. The vorticity flux in the separated flow (Fig. 12a) oscillates about a zero mean during the passage of counter-rotating wake vortices, which corresponds to small fluctuations in the lift force. The application of pulsed modulation results in a sharp positive peak (0.3), which corresponds to the passage of a clockwise vortex shed from the top surface of the airfoil. The positive peak is followed by a negative peak (with a peak level of -0.27) that is associated with the passage of a counter-clockwise vortex corresponding to the re-establishment of lift. Subsequently, the vorticity flux changes sign two more times before reaching low-level oscillations about zero.

When the flow reattaches, the incremental change in the circulation with respect to the baseline flow $-\Delta\langle\hat{\Gamma}\rangle$ (Fig. 12b) initially diminishes to a value of -0.6 (resulting in a momentary decrease in the lift coefficient) and then recovers to a value of 0.46 with the shedding of the second counter-clockwise vortex.

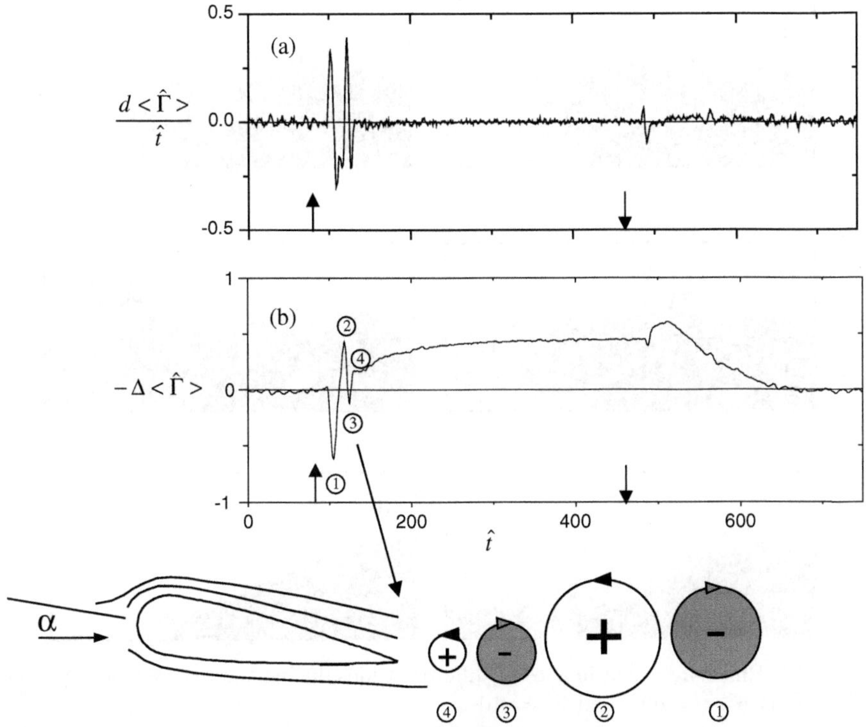

Fig. 12. Phase-averaged vorticity flux (**a**) and circulation increment (**b**) for $\alpha = 17.5°$, $\gamma = 60°$, and $F^+ = 10$

It appears that the shedding of the "starting vortex" causes partial trailing edge separation, which is manifested by the shedding of another (weaker) clockwise vortex followed by a train of vortices of alternating signs (see cartoon). The circulation (and lift coefficient) ultimately converges to its attached value, which is in good agreement with the lift coefficient obtained from the pressure measurements. When the control is turned off the circulation initially increases before settling at the baseline-stalled level, which is similar to the transient variation of lift on a pitching airfoil during dynamic stall.

The flow mechanisms associated with the reattachment process are demonstrated in a sequence of smoke visualization images in Figs. 13a-f (the separated flow, in the absence of control, is shown for reference in Fig. 13a). The smoke is injected in a sheet at the center span and is illuminated using a pulsed laser. At $\hat{t} = t/T = 8$ after the control is activated (Fig. 13b) a vortex with a negative vorticity, which is associated with the shedding of the "trapped" vorticity of the separated flow, is formed near the leading edge. At $\hat{t} = 18$ (Fig. 13c), this vortex has grown in size and is advected downstream. Note the beginning of the formation of a second vortex with a negative vorticity near the leading edge. As time progresses these two vortices continue to

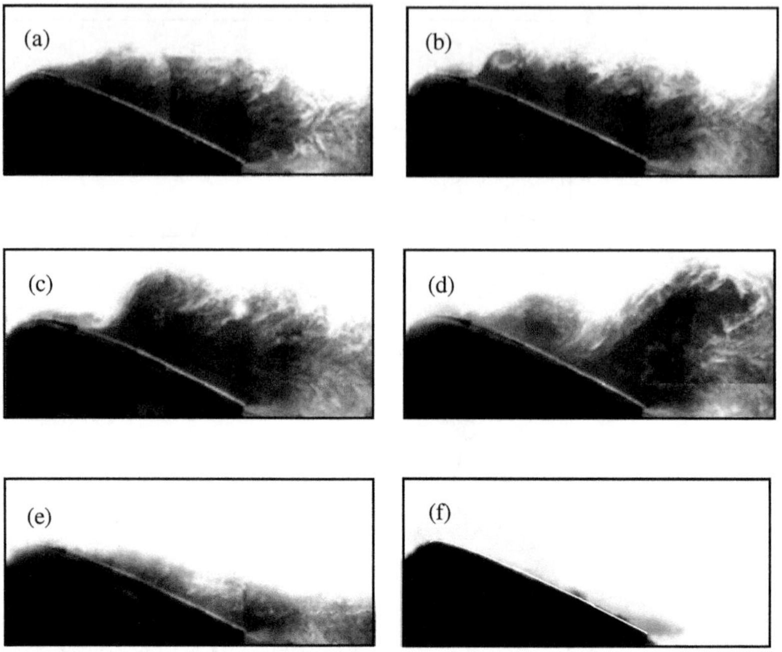

Fig. 13. Phase-averaged images at different times during the reattachment process for $\alpha = 17.5°$, $\gamma = 60°$, and $F^+ = 10$

advect towards the trailing edge while increasing in size (by the time the first vortex reaches the trailing edge its size is more than half of the airfoil's chord, Fig. 13d). At $\hat{t} = 33$ (Fig. 13e) a third vortex is formed, however, this vortex resides near the surface of the airfoil and does not increase in size as the first two vortices. Following the transient ($\hat{t} = 125$, Fig. 13f) the flow is completely attached to the surface of the airfoil and there is no evidence of vortical structures. Note that the vortices with a positive vorticity that contribute to the establishment of a positive lift are not present in these flow visualization images because they are shed from the trailing edge off the window used for the flow visualization.

In the measurements described above, the reduced frequency of the actuator jets is $F^+ = 10$. As noted in Sects. 1 and 3, earlier investigations [4, 10] of separation control have primarily emphasized actuation frequencies $F^+ \sim O(1)$ regardless of the choice of actuators. In order to demonstrate the effect of the actuation frequency on the details of reattachment and on the concomitant variation in lift, a series of experiments are conducted in which the synthetic jets are driven at $F^+ = 0.95$ (71 Hz). Similar to Fig. 11, Fig. 14 shows a color raster plot of the phase-averaged cross-stream distributions of the spanwise vorticity. It is remarkable that the initial transient following the application of the pulse-modulated control are very similar to the measurements shown

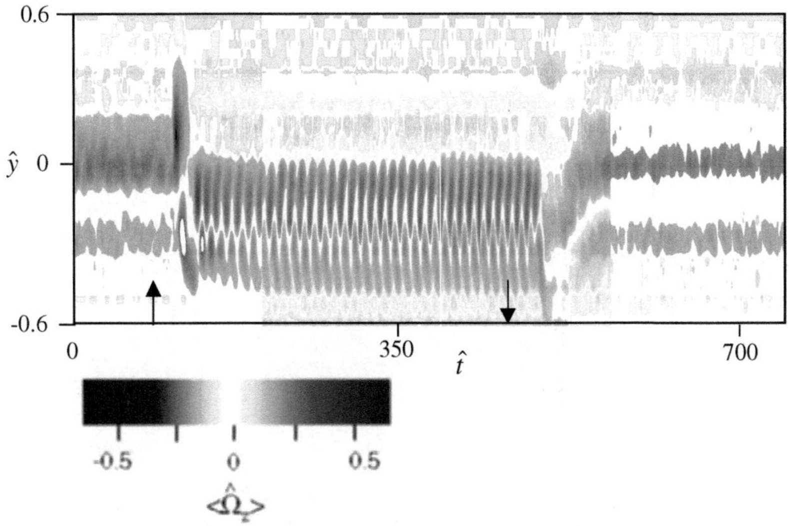

Fig. 14. Phase-averaged color raster plot of the cross-stream distribution of the spanwise vorticity; $F^+ = 0.95$

in Fig. 11. However, in contrast to the reattachment at the higher reduced frequency in which the shedding of organized vortical structures appears to subside following the transient, the reattachment at a reduced frequency of order one appears to be accompanied by the coherent shedding of a train of strong vortices at the actuation frequency. The flow transients associated with the termination of the pulse modulation and the subsequent relaxation of the wake are similar to the corresponding transients at the higher actuation frequency suggesting that the separation process in both cases is similar.

Using the phase-averaged cross-stream distributions of the streamwise velocity and spanwise vorticity, the phase-averaged increment (relative to the baseline flow) of the circulation is estimated and shown in Fig. 15 ($F^+ = 0.95$ and 10 are shown using solid and dashed curves, respectively). When the flow reattachment begins, $-\Delta\langle\hat{\Gamma}\rangle$ exhibits a similar transient at both control frequencies (i.e., a negative peak followed by a positive peak). However, while for high frequency actuation the circulation ultimately reaches a steady level, low frequency actuation results in oscillations of $-\Delta\langle\hat{\Gamma}\rangle$ at the actuation frequency with peak-to-peak fluctuations of up to 45% of the mean level for the attached flow.

6 Pulse-Modulated Reattachment

The flow transients associated with the reattachment and separation processes that are described in Sect. 5 are exploited to further enhance the effectiveness

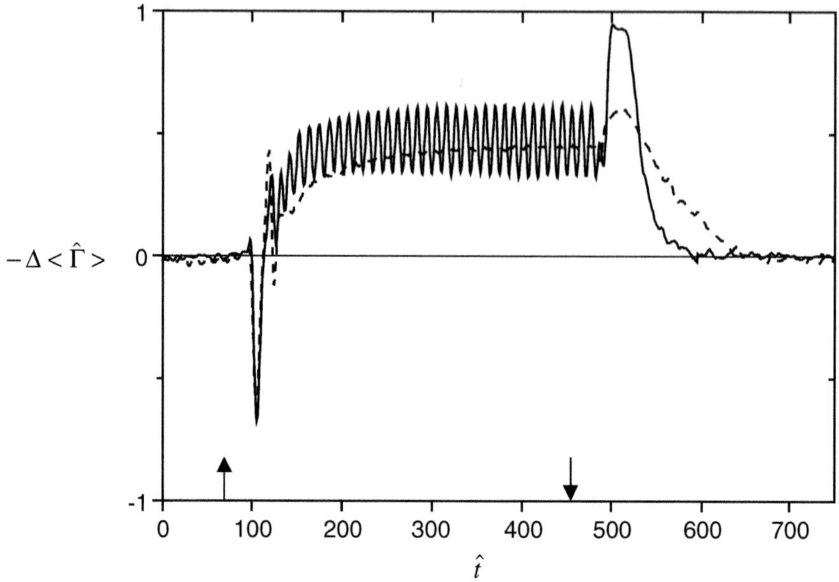

Fig. 15. Phase-averaged circulation increment for $\alpha = 17.5°$ and $\gamma = 60°$. $F^+ = 0.95$ (solid), and 10 (dash)

of the jet actuators. Pulse-modulation of the actuation input, which may be useful in situation when either the streamwise placement or the strength of the jet actuators are sub-optimal, is demonstrated by placing the jet actuator at $\gamma = 42°$, where as shown in the earlier work of Smith et al. [13], C_μ yields some measure of proportional control of the lift coefficient. The effect of C_μ on the distribution of the pressure coefficient around the airfoil (at $F^+ = 10$) is shown in Fig. 16 for $C_\mu = 3.7 \cdot 10^{-3}$ and $4.6 \cdot 10^{-3}$ (open and solid symbols, respectively; the pressure distribution in the absence of control is shown using a solid curve). It is evident that when $C_\mu = 4.6 \cdot 10^{-3}$, the flow is fully attached ($C_L = 0.8$) and a strong low-pressure region is present near the leading edge on the suction side of the airfoil followed by a rapid pressure recovery towards the trailing edge. However, a relatively small (18%) reduction of the momentum coefficient to $C_\mu = 3.75 \cdot 10^{-3}$, results in a partially reattached flow and a substantial degradation of the lift coefficient to $C_L = 0.4$. The pressure distribution exhibits a much smaller suction peak near the leading edge followed by a separation bubble that extends throughout most of the upper surface of the airfoil.

By exploiting the flow transients that are associated with the onset and removal of the actuation (cf. Sect. 5), the performance of the actuators at reduced levels of momentum coefficient, can be substantially enhanced by pulse modulation of their resonance waveform (nominally at $F^+ = 10$). While the period t' and duty cycle $\tilde{d}$ of the modulating pulse train can be independently

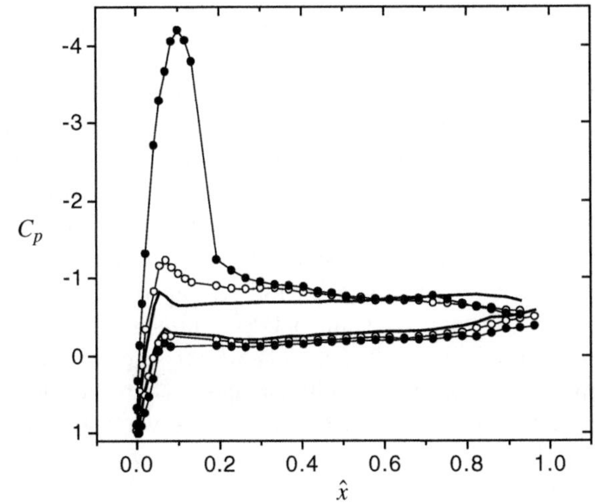

Fig. 16. Pressure coefficient distributions around the airfoil. $C_\mu = 3.75.10^{-3}$ (—•—), $C_\mu = 4.6\ .10^{-3}$ (—o—), and baseline (——)

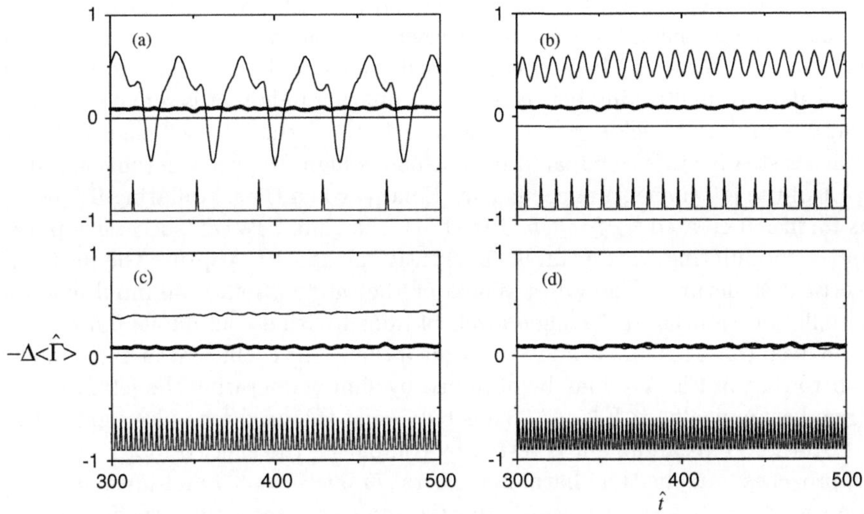

Fig. 17. Phase-averaged circulation increment for $\alpha = 17.5°$ and $\gamma = 42°$. $f^+ = 0.27$ (**a**), 1.1 (**b**), 3.3 (**c**) and 5.0 (**d**)

varied, in what follows $\tilde{d} = 0.25$ and the modulating frequency $f^+ = 1/\text{t'}$ is varied between 0.27 and 5.0.

The long-term variation of the phase-averaged circulation (phase locked to the modulating wave train) with f^+ following the decay of the initial transients is shown in Figs. 17a-d for $f^+ = 0.27$, 1.1, 3.3 and 5.0, respectively (the time trace corresponding to the unmodulated actuation is shown for reference in each plot using symbols). The modulation frequency $f^+ = 0.27$ (Fig. 17a) corresponds to the "natural" passage frequency of the vortices during the initial (transient) stages of the reattachment process (Fig. 12). The resulting circulation exhibits oscillations that are similar in magnitude and duration to the transient stages of the reattachment with shedding of similar vortical structures. The phase of each pulse of the modulating wave train is timed so that it re-triggers reattachment before the flow separates again. This phase is evidently a bit off, because the circulation apparently exhibits low-frequency variations (having a period of the order of 60T). When f^+ is increased to 1.1 the elapsed time between pulses within the modulating wave train is decreased (Fig. 17b) and the large oscillations in the circulation are substantially attenuated. This suggests that the modulating pulses are timed to prevent continuous shedding of large vortical structures and the corresponding variations in circulation. The recovery of an asymptotic circulation of approximately $-\Delta\Gamma = 0.45$ also suggests that the actuation allows the accumulation and maintenance of (clockwise) vorticity on the suction side of the airfoil even though the reattachment is unsteady and the circulation oscillates with peak-to-peak variations of 42% of its asymptotic mean level.

Further increase in f^+ to 3.3 (Fig. 17c) results in a circulation that is similar to the magnitude of the piecewise-averaged circulation in Fig. 17b ($f^+ = 1.1$). However, the absence of oscillations at the modulating frequency indicates optimal timing between the modulating pulses. It is remarkable that pulse modulation yields an increase of $\sim$ 400% in the lift coefficient (when it reaches steady state) compared to continuous high-frequency actuation but at 25% of the jet momentum coefficient. Finally, when the modulating frequency is further increased to $f^+ = 5$ (Fig. 17d), the time between successive pulses of the modulating wave train is apparently too short to capture the unsteady vortical structures. The effectiveness of the modulation is minimal and the circulation returns to the same levels obtained with a continuous pulse train.

Given these results, it might be argued that a lift coefficient increment similar to that of Fig. 17c may be obtained by simply operating the jet actuators time-harmonically at $F^+ = f^+$ thus bypassing the need for pulse modulation altogether. This argument is tested by comparing the effect of both actuation approaches namely, time harmonic excitation at $F^+ = 3.3$ and pulse modulation at $f^+ = 3.3$ *while maintaining the same jet momentum coefficient* C_μ. Time traces of the normalized circulation for both cases are shown in Fig. 18 (time harmonic actuation is plotted using symbols). Also shown are the input waveforms that lead to identical C_μ over one period of the actuation. These data show that time harmonic actuation does not yield the same levels of lift coefficient as pulse modulated actuation.

The flow field above the airfoil is computed from a sequence of PIV images taken in the x-y plane ($z = 0$) that are each comprised of three partially overlapping frames (each consisting from 150 image pairs) measuring 10 cm $\times$ 10 cm ($\alpha = 20°$, $Re_c = 310,000, \gamma = 45°$, and $C_\mu = 4.5 \cdot 10^{-3}$). Cross-stream maps of time-averaged velocity vectors are shown in Figs. 19a-d. In the absence

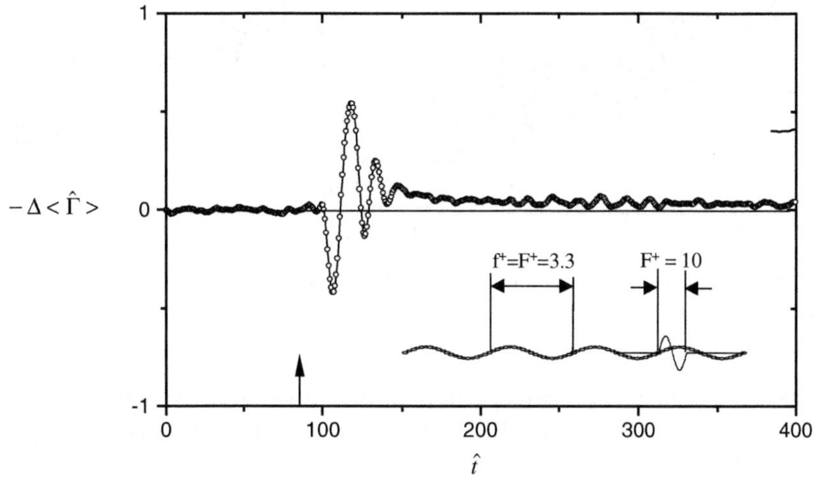

Fig. 18. Phase-averaged circulation increment for $\alpha = 17.5°$ and $\gamma = 42°$. (—o—) continuous periodic actuation and (——) pulsed reattachment technique

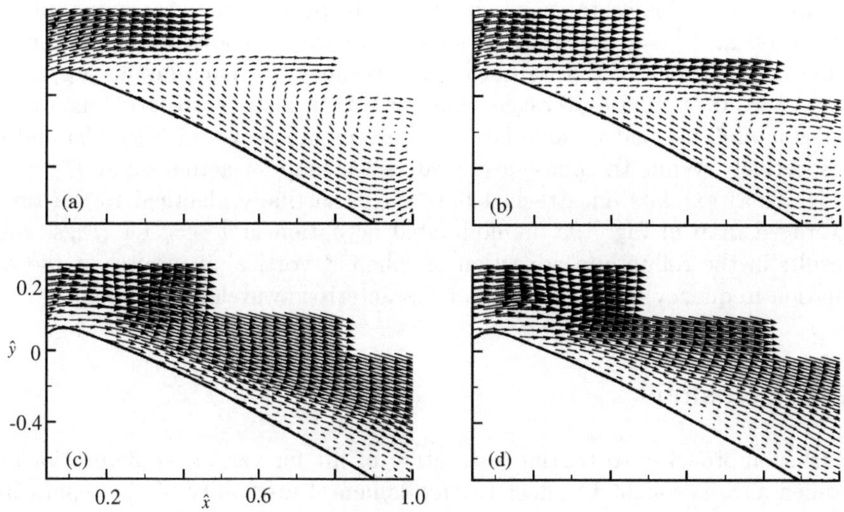

Fig. 19. Time-averaged vector field for $\alpha = 20°$ and $\gamma = 45°$. Baseline (**a**), $F^+ = 10$ (**b**), $f^+ = 1.1$ (**c**), and $F^+ = 1.1$ (**d**)

of control (Fig. 19a), the separated flow exhibits a large recirculating flow domain above the entire upper surface of the airfoil. At this low level of C_μ, unmodulated actuation at $F^+ = 10$ (Fig. 19b) leads to a slight downstream migration (to $\hat{x} \approx 0.3$) of the point of separation and the flow is completely separated thereafter.

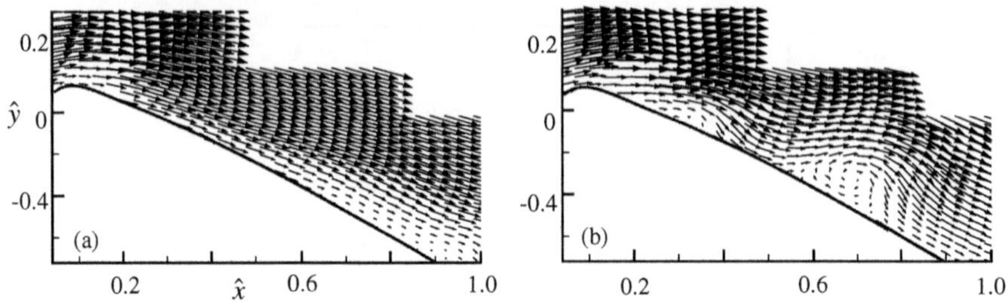

Fig. 20. Phase-average velocity vector fields for $\alpha = 20°$, $\gamma = 45°$; $f^+ = 1.1$ (**a**) and $F^+ = 1.1$(**b**)

When the flow is actuated using pulse modulated actuation at $f^+ = 1.1$ of an $F^+ = 10$ carrier (Fig. 19c), the flow is attached over most of the surface of the airfoil. It is interesting to note that for $\hat{x} > 0.7$, the boundary layer of the attached flow becomes noticeably thicker presumably as a result of a sub-optimal pressure recovery. The vector image suggests that the flow may be separated just upstream of the trailing edge of the airfoil. A comparison of this vector map with the corresponding map for unmodulated actuation at $F^+ = 1.1$ (Fig. 19d) shows that the unmodulated actuation results in thickening of the wall layer that commences much farther upstream (i.e., $\hat{x} > 0.2$). Phase-averaged velocity vector maps corresponding to the pulse modulated and unmodulated actuations are shown in Figs. 20a and b, respectively. While the phase-averaged vector map for actuation at $f^+ = 1.1$ (Figs. 20a) exhibits an attached flow and is virtually identical to the time-averaged map in Fig. 19c, unmodulated actuation at $F^+ = 1.1$ (Figs. 20b) results in the rollup and advection of coherent vortical structures at the actuation frequency having a nominal characteristic wavelength of 0.4c.

7 Conclusions

Earlier approaches to partial restoration of lift for various configurations of stalled airfoils sought to affect the fundamental instability of the separating shear layer and thereby exploit a Coanda-like deflection of that shear layer towards the airfoil surface. This has been typically achieved by using time-periodic (mostly time-harmonic) actuation input having a dimensionless frequency (that nominally scales with the length of the separated flow domain) of order one, which is also within the receptivity of the separating shear layer. The present work explores the potential of bypassing flow separation by direct alteration of the streamwise pressure distribution along the surface that is induced by a local displacement of the cross flow as a result of its interaction with surface-mounted actuator jets. The quasi-steady interaction domain is formed using synthetic (zero net mass flux) jets that are typically operated at frequencies that are at least an order of magnitude higher than the "natural" (or shedding) frequency of the airfoil.

The utility of this actuation approach for the prevention of flow separation over unconventional airfoils where aerodynamic design may be constrained and compromised by other design considerations is investigated on an unconventional symmetric airfoil for which the baseline configuration is normally stalled for angles of attack exceeding 5°. Actuation is effected in the vicinity of the leading edge (on both sides of the stagnation point) at a dimensionless frequency that is at least an order of magnitude larger than the characteristic shedding frequency of the airfoil. The actuation results in fully or partially attached flow with substantial augmentation of lift over a broad range of angles of attack (up to the maximum angle tested, $\alpha = 25°$). Over a smaller range of angles of attack (up to 17.5°), the increases in lift (up to 100% compared to the corresponding baseline) is also accompanied by a decrease in pressure drag (up to 45%).

In the present experiments, the synthetic jet actuators are operated over a range of frequencies up to $F^+ = 20$ (which is about 30 times higher than the natural shedding frequency of the separated flow). While at low actuation frequencies (i.e., $F^+ < 4$) the recovered lift to (pressure) drag ratio L/D_p decreases with increasing F^+, actuation at $F^+ \geq 10$ results in larger L/D_p that appears to be invariant with the actuation frequency. Spectral measurements in the wake of the airfoil indicate that there is a substantial difference in the nature of flow attachment at low and high actuation frequencies. While at low actuation frequencies the spectra are typically dominated by a large spectral component at the actuation frequency, actuation at higher frequencies result in a *featureless spectrum* and the emergence of a spectral band having a $-5/3$ slope indicating enhanced dissipation.

Numerical simulations (using LNS) of the flow over the airfoil yielded excellent agreement with the measured time-averaged surface pressure distributions for the baseline (stalled) flow and in the presence of actuation ($F^+ = 10$ and $C_\mu = 3.5 \times 10^{-3}$). While the stalled flow is dominated by the formation of clockwise (CW) vortices near the leading edge, which are advected into the airfoil's wake, actuation results in a completely attached flow with no discernible shedding of large vortical structures. On the other hand, simulations at low actuation frequencies (e.g., $F^+ = 3.3$) show the time-periodic formation (at the actuation frequency) of a CW vortices that are advected along the surface of the airfoil and appear to be the result of the tilting of the separating shear layer towards the surface. The instantaneous pressure distributions about the surface of the airfoil show that the motion of these vortices is accompanied by suction (low pressure) peaks that are comparable to the instantaneous suction peaks near the leading edge of the airfoil and therefore may contribute to variations in the lift and drag forces on the airfoil.

The flow transients that are associated with flow attachment and separation are investigated using time-modulated control input using phase-locked two-component hot-wire anemometry in the wake of the airfoil. Flow attachment begins with the shedding of a strong clockwise vortex indicating *a momentary reduction in lift* that is followed by a stronger counter-clockwise vortex indicating the re-establishment of lift. The flow transients associated with the application of the actuation are similar at low and high actuation frequencies. However, while at high actuation frequencies the reduced wake of the attached flow ultimately reaches a quasi steady symmetric vorticity distribution, at $F^+ \sim O(1)$ there is coherent shedding of a train of vortices of alternating sign at the actuation frequency and therefore oscillations of the circulation owith peak-to-peak fluctuations of up to 45% of the mean level for the attached flow).

The flow transients that are associated with the onset (or removal) of the actuation are exploited to improve the efficacy of the actuation by pulsed modulation of the excitation input. The pulsed modulation frequency of successive bursts of the driving signal apparently helps to "capture" the vorticity produced during the initial stages of the separation process on the suction side of the airfoil and thus to increase the lift force. The actuator resonance waveform (nominally at $F^+ = 10$) is pulse modulated where the period and duty cycle of the modulating pulse train are independently controlled. In the present work, the (dimensionless) modulation frequency f^+ is varied between 0.27 and 5.0 and the duty cycle (and consequently C_μ) is reduced to 25%. The oscillations in the circulation become minimal when f^+ is 3.3. At higher modulation frequencies, the effectiveness of the modulation is minimal and the lift coefficient has the same level as with continuous high frequency actuation.

Acknowledgements

This work was supported in part by AFOSR (monitored by Drs. T. Beutner, M. Glauser and J. M. McMichael), DARPA, and by The Boeing Company, St. Louis. The authors gratefully acknowledge contributions by Drs. D.R. Smith, V. Kibens, M. Lal and D.E. Parekh.

References

[1] Ahuja, K. K. and Burrin, R. H., "Control of Flow Separation by Sound", AIAA Paper 84-2298, January 1984.

[2] Zaman, K. B. M. Q., Bar-sever, A., and Mangalam, S. M., "Effect of Acoustic Excitation on the Flow over a Low-Re Airfoil" *Journal of Fluid Mechanics,* Vol. 182, 1987, pp. 127–148.

[3] Bar-Sever, A., "Separation control on an airfoil by periodic forcing", *AIAA J.* *27*(6), 820–823, (1989).

[4] Seifert, A., Bachar, T., Koss, D., Shepshelovich, M. and Wygnanski, I., "Oscillatory Blowing: A Tool to Delay Boundary-Layer Separation", *AIAA Journal* *31*(11), 1993, pp. 2052–2060.

[5] Williams, D., Acharya, M., Bernhardt, J., and Yang, P., "The Mechanism of Flow Control on a Cylinder with the Unsteady Bleed Technique", AIAA Paper 91-0039, January 1991.

[6] Huang, L. S., Maestrello, L. and Bryant, T. D., "Separation Control over an Airfoil at High Angles of Attack by Sound Emanating from the Surface", AIAA Paper 87–1261, 1987.

[7] Hsiao, F.-B., Liu, C.-F., and Shyu, J-Y., "Control of Wall-Separated Flow by Internal Acoustic Excitation", *AIAA Journal 28*(8), 1990, pp. 1440–1446.

[8] Hsiao, F.-G., Shyu, R.–N., and Chang, R.-C., "High Angle-of-Attack Airfoil Performance Improvement by Internal Acoustic Excitation.", *AIAA Journal 32*(3), 1994, pp. 655–657.

[9] Chang, R.-C., Hsiao, F.-B., and Shyu, R.-N., "Forcing Level Effects of Internal Acoustic Excitation on the Improvement of Airfoil Performance", *Journal of Aircraft 29*(5), 1992, pp. 823–829.

[10] Seifert, A., Darabi, A., and Wygnanski, I., "Delay of Airfoil Stall by Periodic Excitation", *Journal of Aircraft 33*(4), 1996, pp. 691–698.

[11] Donovan, J. F., Kral, L. D. and Cary, A. W., "Active Flow Control Applied to an Airfoil", AIAA Paper 98–0210, January 1998.

[12] Smith, B. L. and Glezer, A., "The Formation and Evolution of Synthetic Jets", *Physics of Fluids*, Vol. 10, No. 9, 1998.

[13] Smith, D. R., Amitay, M., Kibens, V., Parekh, D. E. and Glezer, A., "Modification of Lifting Body Aerodynamics using Synthetic Jet Actuators", AIAA Paper 98–0209, January 1998.

[14] Amitay, M., Honohan, A., Trautman, M. and Glezer, A., "Modification of the Aerodynamic Characteristics of Bluff Bodies using Fluidic Actuators", AIAA Paper 97–2004, June 1997.

[15] Amitay, M., Smith, B. L. and Glezer, A., "Aerodynamic Flow Control using Synthetic Jet Technology", AIAA Paper 98–0208, January 1998.

[16] Erk, P., "Separation Control on a Post-Stall Airfoil Using Acoustically Generated Perturbations", Ph.D. Dissertation, Technische Universitat Berlin, 1997.

[17] Amitay, M. and Glezer, A., "Aerodynamic Flow Control of a Thick Airfoil Using Synthetic Jet Actuators", FEDSM99-6922, Proceedings of the 3rd ASME/JSME Joint Fluids Engineering Conference, July 18–22, 1999, San Francisco, California.

[18] Amitay, M., Kibens, V., Parekh, D. E. and Glezer, A., 1999, "Flow Reattachment Dynamics over a Thick Airfoil Controlled by Synthetic Jet Actuators", AIAA Paper 99–1001.

[19] Wu, J.-Z., Lu, X.-Y., Denny, A.G., Fan, M. and Wu, J.-M., "Post-stall control on an airfoil by local unsteady forcing", *Journal of Fluid Mechanics*, vol. **371**, pp. 21–58, 1998.

[20] Palaniswamy, S., Metacomp Technologies, AVIA Report, 2000.

Control of Mixing and Reactive Flow Processes

A.R. Karagozian

Department of Mechanical and Aerospace Engineering, University of California, Los Angeles, CA 90095-1597
ark@seas.ucla.edu

Abstract. The interdisciplinary field of reactive flow control is one that holds a great deal of promise for the optimization of complex phenomena occurring in many practical systems, ranging from automobile and gas turbine engines to environmental thermal destruction systems. The fundamental underpinnings of combustion control, however, require a detailed level of understanding of complex reactive flow phenomena, and, in the case of closed-loop active control, require the ability to sense (monitor) and actuate (manipulate) flow processes in a spatially distributed manner in near real time. Hence the ultimate growth and success of the field of reactive flow control is intimately linked: 1) to advances in the understanding, simulation, and model reduction for complex reactive flows, 2) to the development of experimental diagnostic techniques, in particular, to the development of physically robust sensors, and 3) to the development of a framework or frameworks for generation of closed loop control algorithms suitable for unsteady, nonlinear reactive flow systems. The present paper seeks to outline the potential benefits and technical challenges that exist for mixing and combustion control in fundamental as well as practical systems and to identify promising research directions that could help meet these challenges.

1 Introduction

The control of combustion processes in practical systems is often related to the control of the mixing processes in the device (e.g., fuel-air mixing, dilution air-hot product mixing, etc.). Maximizing the degree of mixing is often sought in a number of applications for purposes of enhancing combustion processes, for example, in scramjet engines where supersonic airspeeds entering the combustor require rapid fuel-air mixing and combustion completeness in a relatively short distance [1, 2]. Yet there are also situations in which "maximized" fuel-air mixing does not always produce optimal effects. This can happen, for example, in combustor configurations in which increased fuel-air mixing or partial premixing can actually increase NO_x (nitrogen oxide) emissions due to the competition between the effects of radiative losses from the flame (due to sooting) and (lean) premixing in lowering flame temperatures [3, 4].

Thus in the optimization and eventual *control* of reactive flow processes, it becomes necessary not only to be able to understand the complex underlying physics governing the nonlinear coupling between flowfield and reaction, but to be able to identify sets of conditions which lead to optimal performance and to be able to sense and actuate within the flowfield in order to recover optimal conditions in response to system upsets. Hence the burgeoning field of combustion control and its application to practical systems requires significant advances in the fundamental fields of reactive fluid mechanics and dynamic systems control, as well as in the design and development of physically robust sensors and actuators.

2 Passive Control Methodologies

"Control" of combustion processes in practical systems has a long history, in many cases without being specifically called such. Most systems have employed **passive control** of the combustion process, in which specific configurations for components such as fuel injectors, combustor liners, and flameholders, are used to achieve specific performance goals or to ameliorate undesirable effects for a prescribed range of operating conditions. One example of passive control from the 1950s is the development of methods to control the "screech" combustion instability observed to occur in ramjet engines [5]. The screech phenomenon is caused by vortices, coincident with flame structures, which are shed alternately from the edges of the step or flameholder in the combustor. The vortices excite transverse modes of acoustic oscillation within the cavity due to periodic heat release associated with the generation of hot products. This combustion instability has been controlled in ramjet dump combustor designs by means of flow obstructers or dampers [6], yet an understanding of and ability to control longitudinal-mode combustion instabilities has been sought more recently through active means [7, 8, 9, 10]. Other examples of passive combustion control include swirlers and atomizers for fuel jets in combustion chambers [11], primary and dilution air jet injection for controlling NO_x formation as well as temperature pattern factor [12, 13], and for scramjet engines, transverse fuel injection into supersonic airflow behind a rearward-facing step [14, 15, 16], or shock-induced mixing concepts [17, 18].

3 Passive Control Example: The Lobed Fuel Injector

Many passive mixing and combustion control methodologies employ systems which generate streamwise vorticity in the flowfield [19]. Examples of flowfields which generate streamwise vorticity include the transverse jet, with which there is associated strong streamwise vortical structures [13, 20], and the shock-induced mixing concept [17, 18], whereby streamwise vorticity arises from baroclinic torque generated by the interaction of a shock with a jet containing light gas (fuel). Other examples, such as swirling jets [11], provide further evidence of the ability of streamwise vorticity to assist with molecular mixing as well as flame stabilization.

An example of a passive mixing/combustion control concept explored in recent years by our research group at UCLA is that of the **Lobed Fuel Injector**. The lobed injector, shown schematically in Fig. 1, is a device in which very rapid initial mixing of reactants can occur through streamwise vorticity generation, producing high strain rates which can delay ignition at fuel-air interfaces. The injector consists of two parallel, corrugated plates (which are initially flat) with a ramp angle α representing growth of the corrugation. Fuel is injected from between the plates into coflowing air. Streamwise vorticity is created by the oppositely oriented secondary flows which develop along the sides of each of the lobes; the secondary flows roll up into counter-rotating vortices oriented in the streamwise direction. These vortical structures act to strain fuel-air interfaces, potentially delaying ignition while assisting with molecular mixing of reactants. Further downstream of the vortical structures, the strain field relaxes, potentially allowing ignition to occur in a premixed or partially premixed mode.

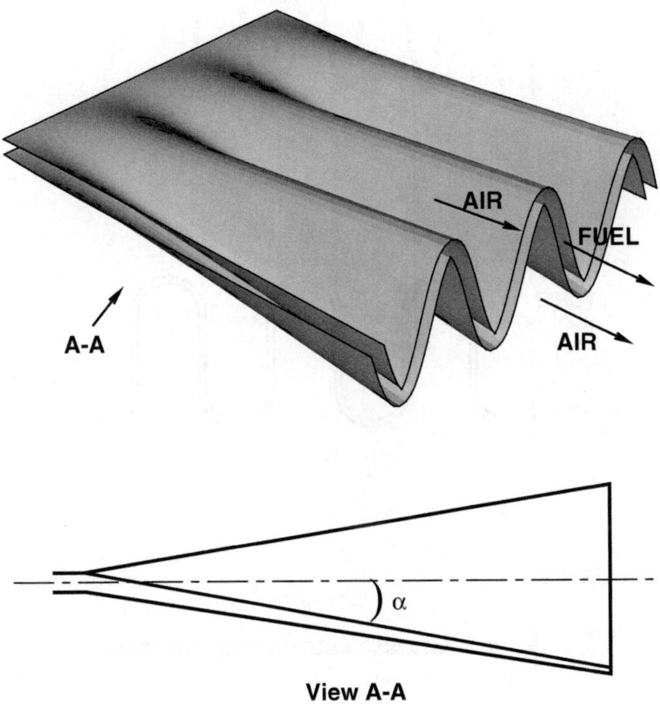

Fig. 1. Schematic of the general lobed injector geometry

The general principle of the lobed geometry has been applied to two-stream, non-reactive, passive mixing control in turbofan engines using a single corrugated plate or interface to mix initially separated fluids [21, 22, 23, 24]. Combustion experiments in a lobed *mixer* geometry, with fuel and oxidizer initially separated by a single lobed splitter plate, demonstrate that the flame spread angle is double that created by a flat splitter plate, indicating enhanced mixing processes and an increased rate of flame propagation [25, 26, 27].

Studies pertaining to the lobed *injector* by our group have included non-reactive mixing experiments, both low speed [28] and transonic [29], combustion experiments with alternative flameholder configurations [30, 31], numerical simulations of lobed injector mixing and reaction processes [32, 33], and fundamental simulations of strained fuel strip processes relevant to the lobed injector concept [34, 35]. Alternative lobe geometries were explored in all of these studies; sample exit plane geometries are shown in Fig. 2, where comparison is generally made with a straight (non-lobed) injector.

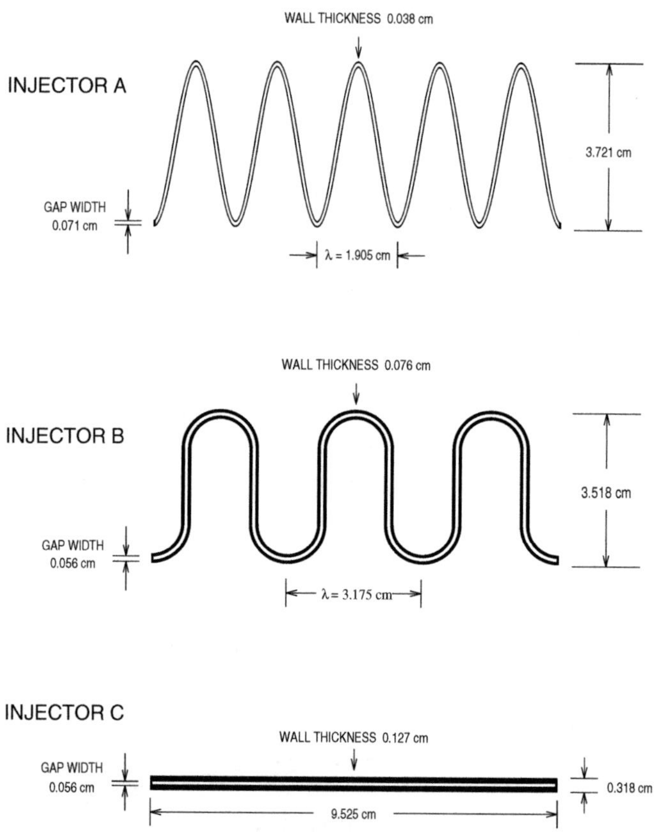

Fig. 2. Comparison of exit plane geometries for three injectors examined in passive combustion control: lobed fuel injectors A and B and straight fuel injector C

Non-reactive mixing experiments [28] employed planar laser-induced fluorescence (PLIF) imaging of acetone seeded with the injectant (carbon dioxide or CO_2 as a fuel "surrogate") to examine the evolution of the flowfield and to quantify local unmixedness and average scalar dissipation rates, from which strain rates were estimated [36]. Molecular mixing was quantified here using the "unmixedness" parameter U [37], which is related to the second moment of the scalar concentration field, enabling comparisons of the local scalar field to be made with the scalar field that would be present if the fluids were completely mixed or completely unmixed. Here the unmixedness U approaches zero for the case of completely molecularly mixed fluids. The local scalar dissipation rate χ [38] was computed as well from the local gradients in mixture fraction [36], then were averaged over the injectant/air interfaces for the fuel injector under examination.

Sample results are shown in Fig. 3 for the evolution of unmixedness for the two lobed injectors (A and B) as compared with the straight fuel injector (injector C). These results suggest that a sharply peaked, sinusoidal injector such as A could have the potential for improved local mixing. Further results, shown in Smith, et al. [28], suggest that lobed injectors produce higher strain or scalar dissipation rates as compared with the non-lobed injector under specific flow conditions.

Yet the implications of this increased local mixing and local and average straining by a lobed configuration do not necessarily always translate directly into lowered NO_x emissions. Results from combustion experiments [30, 31] show dramatic differences in the visual structure of the flames formed in the combustion tunnel, with lobed injectors creating flames which were much bluer and far better mixed, at least locally, than flames created by a straight

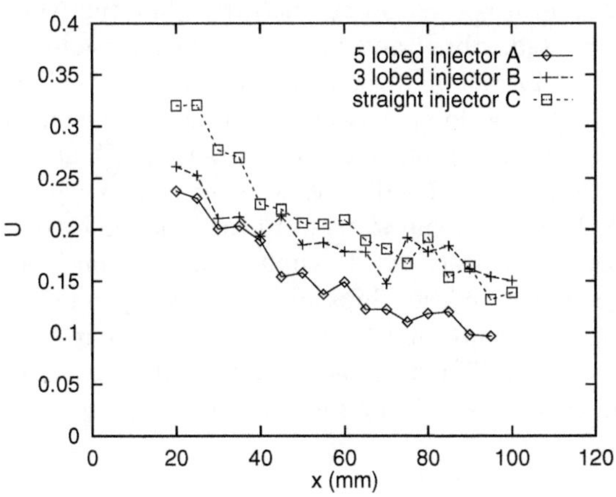

Fig. 3. Lobed injector measurements of local unmixedness U as a function of downstream distance x (in mm) for air flow at 11.5 m/s. Comparisons are made among injectors A, B, and C (results from Smith, et al. [28])

injector. This would imply that the flame structures created by lobed injectors were locally leaner than for non-lobed injectors, which in fact created long, planar diffusion flame-like structures which soot rather heavily. As shown in emissions plots in Fig. 4, there was a very strong dependence of NO (nitric oxide) and overall NO_x emissions from the lobed injectors on the combustion tunnel's fuel-air mass flux (or equivalence) ratio M (also called ϕ_{tunnel}), but with a much lesser dependence on M for emissions from the straight injector. At very lean mixtures, the lobed injector NO_x emissions could be made lower than for the non-lobed injector, but at higher equivalence or mass flux ratios, the strong sooting by the straight injector actually appeared to reduce local flame temperatures due to radiative losses [3, 4], thus reducing thermal NO formation [39] as compared with the lower-sooting lobed injector flames. In addition, at very low equivalence ratios, carbon monoxide or CO emissions could increase (see Fig. 4) since the reaction converting CO to CO_2 is quenched under such conditions [40, 41] due to the flow of cooler air exterior to the fuel-air mixture.

When this exterior airflow was restricted by the placement of confining ceramic wedges about the lobed injectors, as described in Mitchell, et al. [31], significantly leaner local mixtures were formed downstream of flameholders without air bypassing, and both NO_x and CO emissions were simultaneously reduced, to levels below 5 and 1 ppm, respectively, and indicated in Fig. 5. In the case where the wedges extended to create an effective flameholder of length $Z = 7\,\mathrm{cm}$, visual flame structures suggested much greater and more intimate fuel-air mixing achieved prior to flameholding. It was thus possible to operate at higher air speeds (lower ϕ_{tunnel}) so that leaner premixed flames could be stabilized in a distributed fashion along the edge of the step formed by the blocks. As a result of burning in a lean premixed mode (which lowered the local flame temperature), Fig. 5 demonstrates that the NO_x emissions were significantly decreased. Simultaneously, the CO emissions were much lower than CO emissions for the flameholder configuration 1. Confining the air to flow mainly between the lobes virtually eliminated the surrounding blanket of coflowing cold air, thus promoting the oxidation of CO in the cases shown in Fig. 5. Hence operating conditions were identified which simultaneously produced low NO_x and CO emissions from the lobed injectors. No such simultaneous lowering of emissions was possible from the non-lobed injector.

Hence the complex relationships among mixing, temperature, soot, excess air, and flame character must be taken into account when attempting to achieve passive control. This interplay is especially critical for *active* mixing and combustion control, as will be discussed below.

4 Active Control Methodologies

While passive control has been used extensively in combustion systems, **active control** of combustion has had more recent focus, whereby flow and reaction

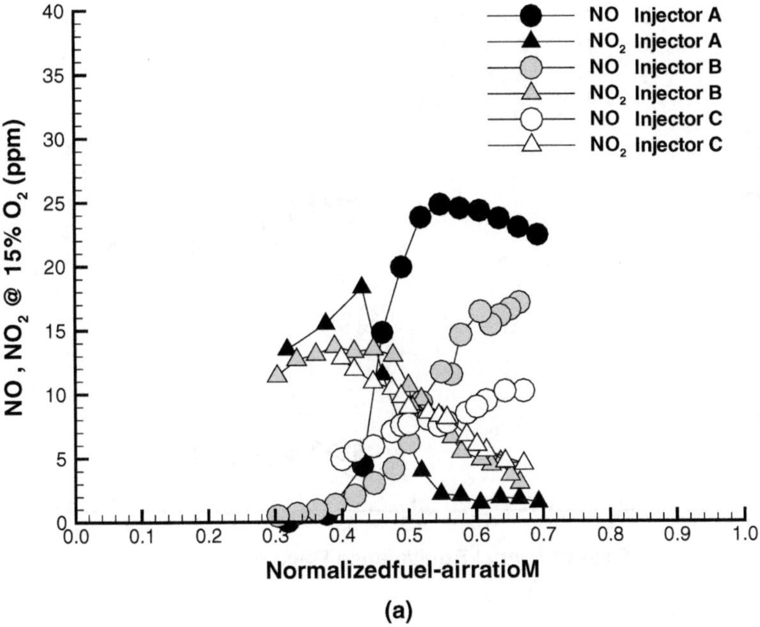

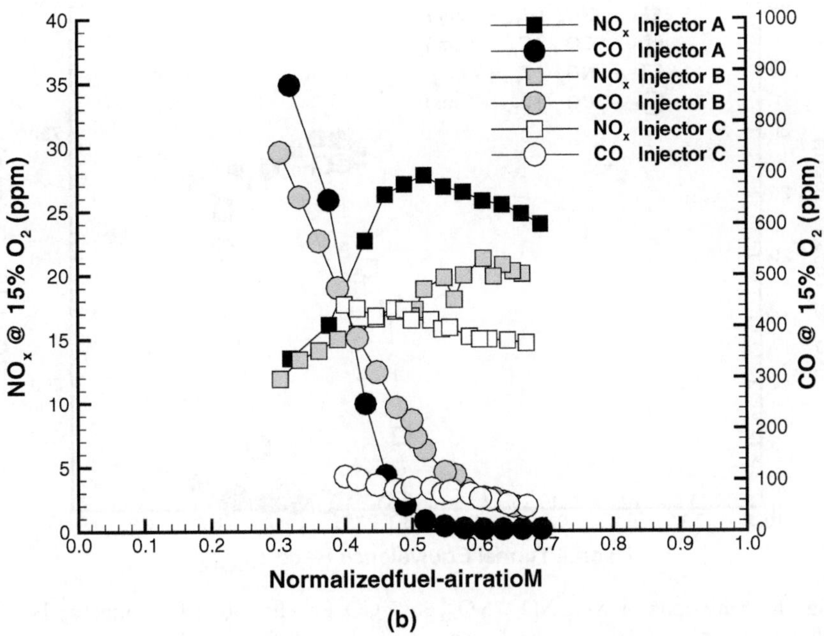

Fig. 4. Emissions of NO, NO_2, NO_x, and CO for the lobed fuel injectors A and B and the straight fuel injector C as a function of normalized tunnel fuel-air mass flux ratio M. The flameholder is situated 7 cm downstream of the injector, and the matched bulk air speeds are 4 m/s (results from Mitchell, et al. [30])

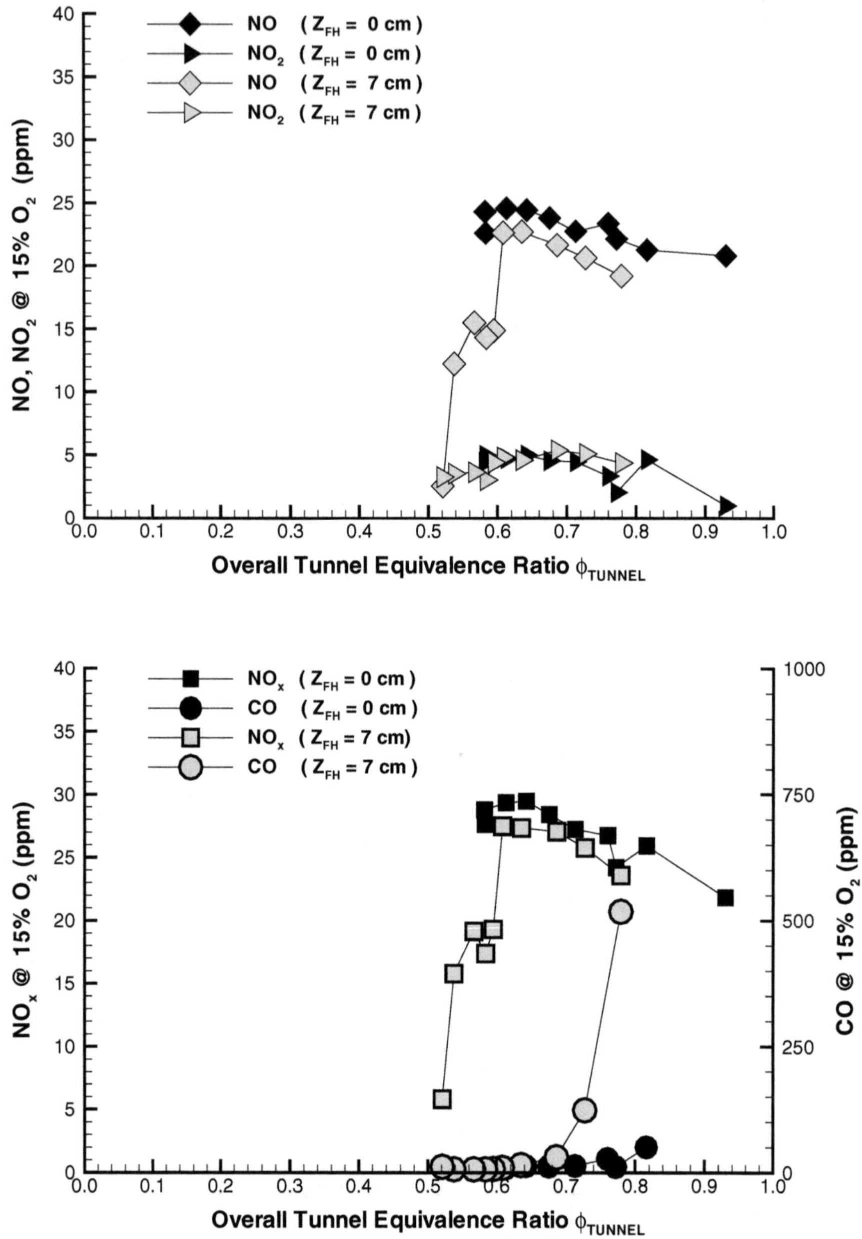

Fig. 5. Emissions of NO, NO_2, NO_x, and CO for the lobed fuel injector B with different lengths of confinement for the ceramic block flameholder arrangement [31] as a function of normalized tunnel fuel-air mass flux ratio $M = \phi_{\text{tunnel}}$. The fuel flow rate was kept constant and the air flow rate was varied to change the value of ϕ_{tunnel}. Here Z_{FH} represents the length of the ceramic block flameholder beyond the extent of the fuel injector exit plane. Larger Z_{FH} values allow greater fuel-air mixing to occur prior to ignition and flameholding

processes are externally acted upon (by an "actuator") as a function of time and flow conditions. Active control is particularly well suited to deal with physical processes in which oscillations (e.g., resulting from acoustically driven combustion instabilities) can occur. Active control in combustion systems is often "open loop", in which the operation of the actuator is not modified in real time in relation to the evolution of the flow or reaction process. More recently, however, "closed loop" control has been explored, whereby sensors placed within the flowfield (generally in cooler sections of the combustor) feed back information to the actuators such that the actuators' operation is altered.

The problem of the closed loop control of combustion processes is especially challenging, even in simple configurations. Because of the high temperatures generated by the combustion process, realistic actuators can operate only within specific regions of the flow (e.g., in the fuel line or near the orifice of a fuel jet), and currently practical sensors (e.g., flush-mounted pressure transducers) can be placed only in relatively cool regions or at the exhaust of the combustion chamber. Consequently, there is a relatively large time lag between the time at which the actuator modifies the flow and the time at which the sensor measures the effect of this action on the reactive flowfield. During this time lag, unsteady mixing and combustion chemistry are dominated by nonlinear effects.

Most active control methods (open or closed loop) have been directed at 1) reducing large amplitude pressure oscillations arising from combustion instabilities in a reactive flow system [42], 2) reducing emissions such as NO_x [39], 3) improving engine performance, e.g., via control of the temperature pattern factor upstream of the turbine blades [43, 44, 44, 45], or all of the above [47]. Examples of open loop methods for control of combustion instabilities include periodic forcing of inlet boundary layers associated with the generation of instability in a dump combustor configuration [10, 48] and control of exit pressure amplitude for a fixed frequency of oscillation [49], while open loop acoustic control of dilution air jets has been demonstrated to reduce NO_x emissions as well as optimize temperature pattern factor [43, 44, 44, 45].

Closed loop control methods vary widely according to the actuation and sensing methods employed and the goals of the control scheme. Closed loop control strategies can be developed on the basis of a reduced order (simplified) model for the reactive flow process [50, 51, 52, 53, 54], or from system identification via input/output data derived from an experiment or from a detailed model [55, 56]. Closed loop control of pulsating combustion in ducted flame configurations (such as the Rijke tube) has been demonstrated using a loudspeaker as the actuator and microphones and/or photomultiplier tubes (PMTs) as sensors to measure sound pressure level and flame emissions, respectively [51, 57, 58, 59, 60]. Poinsot, et al. [9] implement a "self-calibrating" fast response adaptive controller with both PMT and microphone sensing, with considerable success in suppressing combustion instabilities in a large scale combustor. Control of acoustically driven combustion instabilities in dump combustors has been explored using upstream boundary layer forcing and then sensing via pressure transducers and photodiode arrays for emissions measurements [53] and by using separately pulsed air and/or fuel injection with pressure transducers for sensing [52, 61]. More recently, closed loop control of combustion instabilities and emissions in gas turbine combustors has been achieved using microphones and OH emission detection for sensing [62].

5 Active Control Example: The Acoustically Resonant Dump Combustor

Pulsating combustion does have benefits in practical systems, and can be exploited to improve burner performance due to quantifiable reductions in emissions (NO_x and CO) and soot, particularly with fuel staging [63, 64, 65]. There has been the suggestion that pulsating combustion increases the convective heat transfer rates as well as turbulence intensity in the flowfield [63]. It is also felt that lower NO_x (thermal NO) emissions can result from periodic rapid mixing of hot combustion gases with cooler residual gases in a highly strained flowfield, with a resulting reduction in residence times of gases at higher temperatures [64, 65]. The following example of an actively controlled, acoustically driven dump combustor demonstrates the potential benefits of control of such pulsating combustion processes.

An active mixing/combustion control concept (open loop) explored in recent years by our research group at UCLA is that of an **acoustically driven, two-dimensional dump combustor** which has been examined for its application to thermal waste destruction [66, 67, 68, 69, 70, 71, 72, 73, 74, 75]. The general configuration for the dump combustor is shown in Fig. 6, including an expanded view of flow/reaction processes in the combustion cavity.

In the dump combustor experiments, propane (C_3H_8) and air at room temperature were introduced into the plenum/mixing section of the combustor, then were accelerated through an inlet section before entering the combustion cavity at the sudden expansion or "dump plane". Premixed flames were then stabilized due to the formation of high temperature recirculation regions downstream of the dump plane [76], but as noted above, vortex shedding coincident with the flames did occur on occation under conditions of natural acoustic excitation [66, 67, 68, 69, 70, 71]. Quartz windows bounded each end of the system in the spanwise direction, allowing appropriate optical access; additional quartz window slits were installed in the side walls to allow the introduction of a sheet of laser light for optical diagnostics [68, 74, 75].

A set of movable ceramic plugs defined the outlet section of this combustor. Gaseous waste surrogates, liquid waste surrogates, or pyrolysis off-gas surrogates were introduced through injectors embedded in these plugs as indicated in Fig. 6. Injection into the recirculation zones allowed surrogates to be trapped for relatively long residence times under potentially high temperature and/or oxygen rich conditions (depending on the inlet gas mixture) so that they could be destroyed more efficiently. Acoustic data were taken

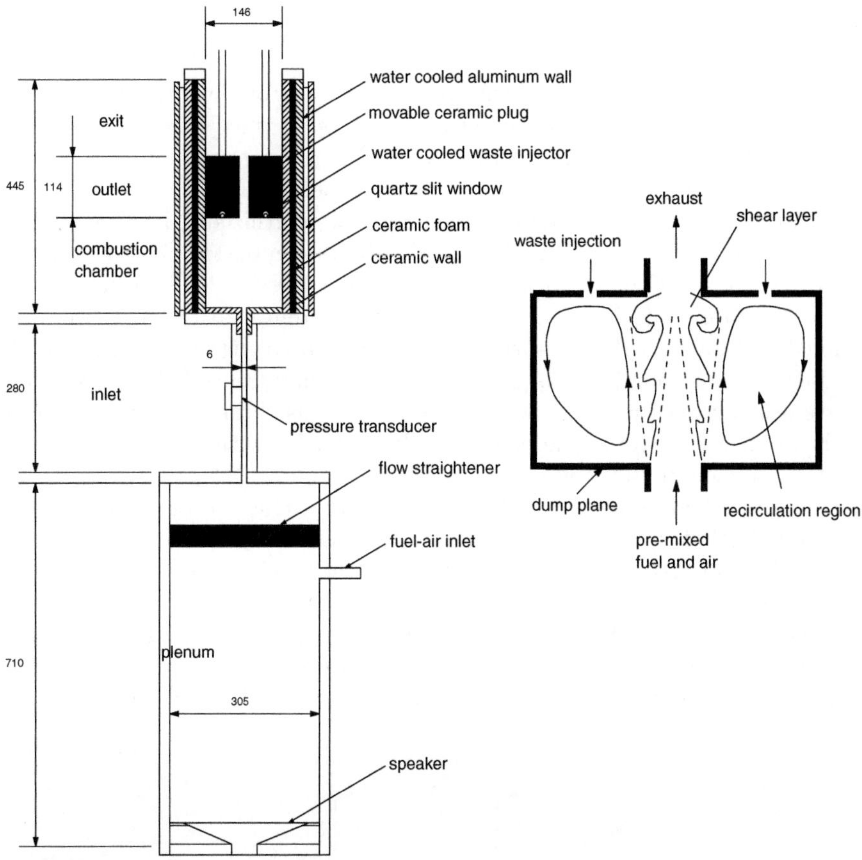

Fig. 6. Schematic of the acoustically resonant dump combustor, including features of the combustion chamber cavity. Dimensions are given in units of mm

using pressure transducers located in the plenum and inlet, and were used for phase-locked imaging [74, 75] While certain operating conditions could produce naturally-occurring acoustic resonances in the device [66], external acoustical forcing could also be imposed using a loudspeaker situated at the bottom of the plenum section [72, 73]. Using a signal generator it was possible to sweep through a range of input forcing frequencies from 0 to 1000 Hz, with forcing amplitude variation up to 150 dB. Thus it became possible to force the loudspeaker at frequencies corresponding to the natural ("on-resonant") modes of the device as well as other ("off-resonant") frequencies.

Performance of the combustor as a potential thermal waste destruction system was quantified in terms of the "destruction and removal efficiency" or DRE for the particular waste surrogate examined, where the DRE is defined as

$$\text{DRE} \equiv \frac{(\dot{m}_{in} - \dot{m}_{out})100\%}{\dot{m}_{in}} \quad (1)$$

and where $\dot{m}_{in}$ and $\dot{m}_{out}$ are the mass flow rates of the principal organic hazardous constituent (waste) entering and leaving the system, respectively. The U.S. Environmental Protection Agency requires a minimum of 99.99% (or "four nines") DRE for hazardous waste incinerators [77].

A variety of different diagnostic techniques were utilized to interrogate flow and reactive processes taking place in the dump combustor. The diagnostics performed included phase-locked hydroxyl radical (OH) PLIF for flame zone characterization [68, 71], NO PLIF from which temperature fields were extracted and quantified [74], particle image velocimetry (PIV) for 2D velocity field characterization [68, 75], and OH* chemiluminescence [72]. Detection of waste destruction the device was made using a gas chromatograph equipped with an electron capture detector or, alternatively, with a flame ionization detector, depending on the waste surrogate examined. Waste surrogates included gases (sulfur hexafluoride or SF_6, methyl chloride or CH_3Cl, and a mixture of ethylene (C_2H_4), benzene (C_6H_6), and nitrogen (N_2)) and liquids (acetonitrile or CH_3CN), each with differing thermodynamic mechanisms for destruction.

The effects of external acoustic actuation on waste destruction and NO_x emissions in the device were found to be significant. For example, Fig. 7 shows that, for certain ranges of forcing frequencies (around 300 Hz, between 500

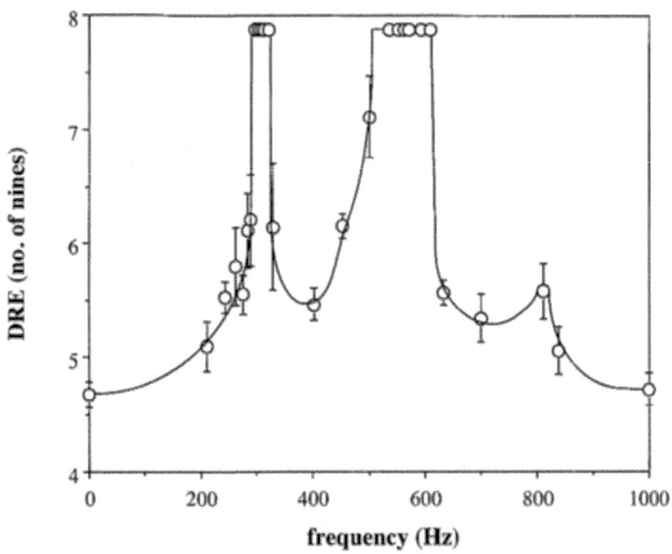

Fig. 7. Destruction and Removal Efficiencies (DREs) of the gaseous waste surrogate SF_6 in the resonant dump combustor during externally forced acoustical excitation for the case where inlet equivalence ratio $\phi = 0.83$. (results from Pont, et al. [72])

and 600 Hz, and around 800 Hz, coincident with the system natural modes [66]), marked improvement in DREs for the waste surrogate SF_6 was achieved. When the device was forced at frequencies other than resonant modes, some improvement over the non-forced combustor performance was achieved, but not nearly as dramatically as at the resonant modes. Destruction of other waste surrogates produced similar levels of improvement during external on-resonance forcing. OH* chemiluminescence as well as NO PLIF images indicated that the enhancement of waste destruction at certain natural modes coincided with flame/core region shortening and broadening due to vortex shedding during the acoustical excitation. Thermocouple as well as NO PLIF quantification of temperature fields in the combustor [73, 74] suggested that the temperatures in the core and recirculation zones (into which the waste surrogates were injected) rose by an average of 30–50 degrees during on-resonance external acoustic excitation; there was no appreciable change in temperatures in these regions during off-resonance external excitation. These observations were consistent with the destruction of SF_6, a surrogate which is pyrolyzed and thus which is sensitive to temperature. Velocity fluctuations as well as inlet core/jet spread were also observed to increase during on-resonant forcing, with little appreciable change during off-resonant forcing [72, 75].

Interestingly, the NO emissions from the combustor simultaneously dropped, in some cases by more than 50%, during external excitation at these same resonant modes, as shown in Fig. 8. These results appear to be consistent with the observations of Keller, et al. [64] who suggest that NO reduction during acoustical excitation in general occurs due to shorter gas residence times at higher temperatures. That the present NO emissions dropped even further at natural frequencies could indicate enhanced residual gas mixing as well at these frequencies.

While closed loop control in this device could provide further improvements in combustor/incinerator performance, and in particular, recovery from upset conditions, current sensing technologies generally limit the location of detection to the regions upstream of flame structures, outside of the combustion zone. Future advances in sensor technologies that could significantly impact combustion control will be identified in the section below.

6 Future Directions

It is clear from the foregoing examples that reactive flow control (passive and especially active control) holds a great deal of promise for improved performance in practical combustion systems. Closed loop active control is particularly challenging in that it requires utilization of sensors (and, to a lesser extent, actuators) that are capable of operating in severe environments and development of robust control algorithms that can be applied over a wide range of operating conditions.

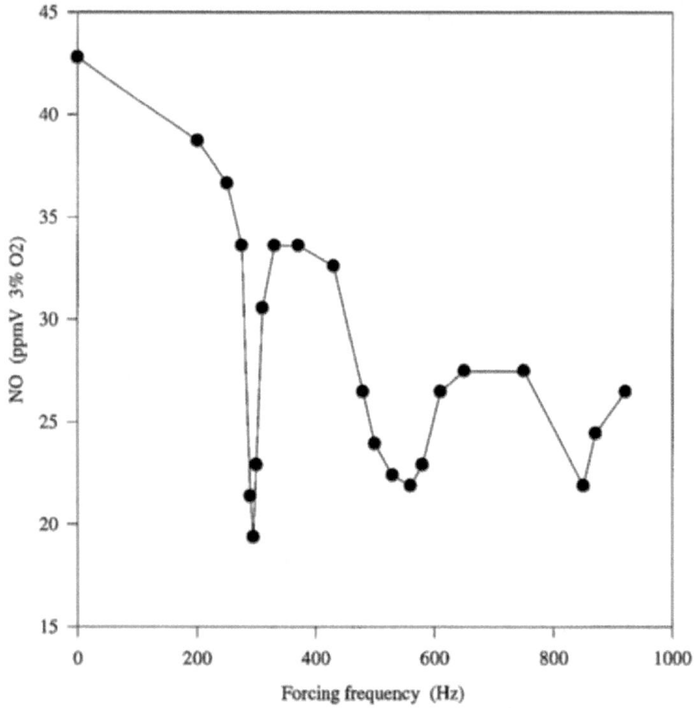

Fig. 8. Nitric oxide (NO) emissions in the dump combustor during externally forced acoustical excitation for the case where inlet equivalence ratio $\phi = 0.83$, inlet velocity $U_i = 4.7\,\mathrm{m/sec}$, and cavity length $= 10.2\,\mathrm{cm}$. No surrogate is injected into the recirculation zones here. Zero forcing frequency refers to the unforced case. (results from Pont, et al. [73])

There are a number of promising **sensor** technologies which are being developed for and implemented in combustion systems. As noted above, operation of these sensors in the severe combustor environment (high temperatures and pressures) becomes a key requirement. MEMS-based sensors are receiving a great deal of attention for applications in non-reactive flow control [78], yet since these are largely silicon-based, their direct applicability in a combustion environment is limited. Aiming at future advances which will allow incorporation of more refractory materials in MEMS devices, Bowman [79] has proposed a microscale chemical sensor for time-resolved NO_x emissions detection. Wireless, ceramic-based RF sensor technology currently under development at UCLA [80] well could be the key to eventual implementation of such flush-mounted sensors (which could detect pressure, species, temperature, etc.) in practical combustors.

Since most critical flow and reaction processes take place away from solid surfaces in a combustion chamber, optical diagnostic methods have become essential interrogation tools for fundamental combustion experiments. Their evolution as real-time sensors also holds great promise. Hanson [81, 82] has developed a diode laser-based system for detection of NO and has implemented the concept in a Navy combustor/incinerator. Fiberoptic sensors have also been used for OH emissions detection and have been incorporated successfully into a closed loop control system for a laboratory gas turbine combustor [83]. The practicality of these optical methods for long cycle time use in an operating engine remains to be seen.

The successful design of a **closed-loop combustion controller** is possible when a model is identified which is able to reasonably capture the dynamics of the flow *and* combustion processes. The controller dimension is an important parameter in any engineering application because of the amount of hardware and computer power necessary to compute a real-time control law. Modern synthesis methods produce controllers with dimensions on the order of the process model. Thus, it is essential, from detailed reactive flow models, to derive reliable, reduced-order models of the combustion process for controller evolution. At present, relatively low order models for combustion processes (e.g., which globally represent transient heat release and pressure perturbations) have been developed and successfully implemented into real-time controllers [47, 51, 52, 53, 54].

Benefits for future combustion controller development may be derived from concepts being employed for non-reactive flow control, which has recently moved in the direction of developing hierarchies of models of differing complexity. These approaches yield insights into methods that may be devised for combustion control. Among the approaches for developing a reduced-order model for non-reactive fluid flow control are: 1) transformation of the governing (Navier-Stokes) equations into a set of ordinary differential equations using, for example, vortex methods [84, 85], Galerkin's methods [86, 87], or proper orthogonal decomposition [88, 89], and 2) system identification [56, 90]. As suggested by Cortelezzi [85, 91], the process of transferring the control strategy from a low-dimensional model to a higher or infinite dimensional model, and eventually to a real flow, should be guided and supported by a dynamical system and time series analysis. In general, the control strategy should be transferable from a low-order model to the next high-order model, provided that the phase space of the two models are topologically equivalent. This framework has been successfully applied to the linear feedback control of near-wall turbulence [86, 87] and to the nonlinear feedback control of the wake past a plate perpendicular to the free-stream [91]. An analogous framework for hierarchies of combustion models could have similar benefits so that, in conjunction with the utilization of high temperature sensors, robust combustion control in a wide range of systems could be a practical reality.

Acknowledgments

The author wishes to acknowledge collaborations and extensive conversations over the years with a number of individuals, most especially with Professor Owen Smith of UCLA, Professor Frank Marble of Caltech, Professor Luca Cortelezzi of McGill University, Professor Sebastien Candel of Ecole Centrale de Paris, and with a large number of UCLA graduate students and postdocs whose efforts are reported here. The UCLA research projects described herein have been supported over the years by NASA, the Office of Naval Research, and the National Science Foundation.

References

[1] Northam, G. B. and Anderson, G. Y., AIAA Paper 86–0159, January, 1986.
[2] Schetz, J. A. and Billig, F. S., AIAA Paper 87–2161, July, 1987.
[3] Turns, S. R., Myhr, F. H., Bandaru, R. V., and Maund, E. R., *Combustion and Flame*, Vol. 93, 1993, pp. 255–269.
[4] Gore, J. P. and Zhan, N. J., *Combustion and Flame*, Vol. 105, 1996, pp. 414–427.
[5] Rogers, D.E., and Marble, F.E., *Jet Propulsion*, 1956, p. 456.
[6] Zukoski, E. E., "Afterburners", Chap. 2 of **Aerothermodynamics of Aircraft Engine Components**, G. C. Oates, Ed., AIAA Educ. Series, 1985.
[7] Yang, V., and Culick, F.E., *Combustion Science and Technology*, Vol. 45, p. 1, 1986.
[8] Sterling, J. D. and Zukoski, E. E., *Combustion Science and Technology*, Vol. 77, No. 4–6, pp. 225–238, 1991.
[9] Poinsot, T., Veynante, D., Bourienne, F., and Candel, S., *Journal of Propulsion and Power*, Vol. 5, 1989, pp. 14–20.
[10] Schadow, K. C., Gutmark, E., and Wilson, K. J., *Combustion Science and Technology*, Vol. 81, 1992, pp. 285–300.
[11] Beer, J. M. and Chigier, N., **Combustion Aerodynamics**, Krieger Publishers, 1983.
[12] Holdeman, J. D., *Progress in Energy and Combustion Science*, Vol. 19, No. 1, pp. 31–70, 1993.
[13] Kamotani, Y. and Greber, I., *AIAA Journal*, Vol. 10, 1972, pp. 1425–1429.
[14] Hollo, S. D., McDaniel, J. C., and Hartfield, R. J. Jr., *AIAA Journal*, Vol. 32, 1994, pp. 528–534.
[15] Wang, K. C., Smith, O. I., and Karagozian, A. R., *AIAA Journal*, Vol. 33, No. 12, 1995, pp. 2259–2263.
[16] Karagozian, A. R., Wang, K. C., Le, A.-T., and Smith, O. I., *J. Propul. Power*, Vol. 12, No. 6, pp. 1129–1136, 1996.
[17] Marble, F. E., Hendricks, G. J., and Zukoski, E. E., AIAA Paper No. 87–1880, 1987.
[18] Ton, V., Karagozian, A. R., Marble, F. E., Osher, S. J., and Engquist, B. E., *Theoretical and Computational Fluid Dynamics*, Vol. 6, 1994, pp. 161–179.
[19] Roshko, A., "The Mixing Transition in Free Shear Flows", in *The Global Geometry of Turbulence*, J. Jimenez, ed., Plenum Press, NY, 1991.

[20] Broadwell, J. E. and Breidenthal, R. E., *Journal of Fluid Mechanics*, Vol. 148, 1984, pp. 405–412.
[21] Tillman, T. G., Paterson, R. W., and Presz, W. M., AIAA Paper No. 87–0610, 1987.
[22] W. M. Presz, B. L. Morin, and R. G. Gousy, *Journal of Propulsion and Power*, Vol. 4, No. 4, 1988, pp. 350–355.
[23] W. A. Eckerle, H. Sheibani, and J. Awad, *Journal of Engineering for Gas Turbines and Power*, Vol. 114, No. 1, 1992, pp. 63–71.
[24] S. A. Skebe, R. W. Paterson, and T. J. Barber, AIAA Paper No. 88–3785-CP, 1988.
[25] McVey, J. B., *Combustion Science and Technology*, Vol. 60, 1988, pp. 447–451.
[26] McVey, J. B. and Kennedy, J. B., AIAA Paper No. 89–0619, 1989.
[27] Waitz, I. A. and Underwood, D. S., *Journal of Propulsion and Power*, Vol. 12, No. 4, 1995, pp. 638–645.
[28] Smith, L. L., A. J. Majamaki, I. T. Lam, O. Delabroy, A. R. Karagozian, F. E. Marble, and O. I. Smith, *Physics of Fluids*, Vol. 9, No. 3, 1997, pp. 667–678.
[29] Majamaki, A. J., Smith, O. I., and Karagozian, A. R., *AIAA Journal*, Vol. 41, No. 4, 2003, pp. 623–632.
[30] Mitchell, M. G., Smith, L. L., Karagozian, A. R., and Smith, O. I., *Twenty-seventh Symposium (Intl.) on Combustion*, The Combustion Institute, Pittsburgh, PA, 1998, pp. 1825–1831.
[31] Mitchell, M. G., Smith, O. I., Karagozian, A. R., *AIAA Journal*, Vol. 42, No. 1, 2004, pp. 61–69.
[32] Strickland, J. H., Selerland, T., and Karagozian, A. R., *Physics of Fluids*, Vol. 10, No. 11, 1998, pp. 2950–2964.
[33] Selerland, T. and Karagozian, A. R., "Numerical Simulation of Reactive Flows Associated with a Lobed Fuel Injector", Paper 97F-108, Western States Section/The Combustion Institute Fall Meeting, 1997.
[34] Gerk, T. J. and Karagozian, A. R., *Twenty-Sixth Symposium (International) on Combustion*, The Combustion Institute, Pittsburgh, PA, 1996, pp. 1095–1102.
[35] Selerland, T. and Karagozian, A. R., *Combustion Science and Technology*, Vol. 131, No. 1–6, 1998, pp. 251–276.
[36] Bish, E. S. and Dahm, W. J. A., *Combustion and Flame*, Vol. 100, 1994, pp. 457–466.
[37] Dimotakis, P. and Miller, *Physics of Fluids A*, Vol. 2, No. 11, 1990, pp. 1919–1920.
[38] Peters, N., *Combustion Science and Technology*, Vol. 30, 1983, pp. 1–17.
[39] Bowman, C. T., *Twenty-fourth Symposium (International) on Combustion*, The Combustion Institute, Pittsburgh, PA, 1992, pp. 859–878.
[40] Lefebvre, A. H., *Gas Turbine Combustion*, Hemisphere Publishing Corp., New York, NY, 1983.
[41] Westbrook, C. K. and Dryer, F. L, *Progress in Energy and Combustion Science*, Vol. 10, 1984, pp. 1–57.
[42] Putnam, A. A., *Combustion-Driven Oscillations in Industry*, American Elsevier, New York (1971).
[43] Vermeulen, P. J., Odgers, J., and Ramesh, V., *J. Eng. Power*, Vol. 104, 1982, pp. 844–852.
[44] Vermeulen, P. J., Ramesh, V., and Yu, W. K., *J. Eng. for Gas Turb. Power*, Paper 86-GT-86, 1986.

[45] Vermeulen, P. J., Ramesh, V., Sanders, B., and Odgers, J., *Journal of Propulsion and Power* Vol. 11, 1995, pp. 261–267.
[46] Vermeulen, P. J. and Ramesh, V., *J. Eng. for Gas Turb. Power*, Vol. 119, No. 3, 1997, pp. 559–565.
[47] McManus, K. R., Poinsot, T., and Candel, S. M., *Progress in Energy and Combustion Science* Vol. 19, 1993, pp. 1–12.
[48] McManus, K. R., Vandsburger, U., and Bowman, C. T., *Combustion and Flame*, Vol. 82, 1990, pp. 75–92.
[49] Najm, H. M. and Ghoniem, A. F., *Journal of Propulsion and Power*, Vol. 10, 1994, pp. 254–262.
[50] Peraccio, A. and Proscia, W., "Nonlinear Heat Release/Acoustic Model for Thermoacoustic Instability", presented at the *Workshop on Dynamics and Control of Combustion Instabilities in Propulsion and Power Systems*, Caltech, Nov. 20–22, 1997.
[51] Annaswamy, A. M. and Ghoniem, A. F., *IEEE Control Systems Magazine*, Vol. 15, 1995, pp. 49–63.
[52] Knoop, P., Culick, F.E.C., and Zukoski, E.E., *Combustion Science and Technology*, Vol. 123, No. 1–6, 1997, pp. 363–376.
[53] Padmanabhan, K. T., Bowman, C. T., and Powell, J. D., *IEEE Trans. Cont. Sys. Tech.*, Vol. 4, pp. 217–229, 1996.
[54] Dowling, A. P., *Journal of Fluid Mechanics*, Vol. 346, 1997, pp. 271–290.
[55] Jacobson, C, "System Identification for Models of Lean Premix Combustion Instabilities", presented at the *Workshop on Dynamics and Control of Combustion Instabilities in Propulsion and Power Systems*, Caltech, Nov. 20–22, 1997.
[56] Murray, R.M., Jacobson, C.A., Casas, R., Khibnik, A.I., Johnson Jr., C.R., Bitmead, R., Peracchio, A.A. and Proscia , W.M., "System Identification for Limit Cycling Systems: A Case Study for Combustion Instabilities", 1998 American Control Conference.
[57] Dines, P. J., "Active Control of Flame Noise", Ph.D. Thesis, Cambridge University, 1984.
[58] Heckl, M. A., "Heat Sources in Acoustic Resonators", Ph.D. Thesis, Cambridge University, 1985.
[59] Gulati, A. and Mani, R., *Journal of Propulsion and Power*, Vol. 8, 1992, pp. 1109–1115.
[60] Gleis, S., Vortmeyer, D. and Rau, W., *AGARD Propulsion and Energetics Panel, 75th Symposium*, 1990, pp. 1–6.
[61] Wilson, K. J., Gutmark, E., Schadow, K. C., and Smith, R. A., *Journal of Propulsion and Power*, Vol. 11, No. 2, 1995, pp. 268–274.
[62] Gutmark, E., Paschereit, C. O., Weinstein, W., and Paikert, B., "Combustion Control in Swirl-Stabilized Gas Turbine Combustors", presented at the *Workshop on Dynamics and Control of Combustion Instabilities in Propulsion and Power Systems*, Caltech, Nov. 20–22, 1997.
[63] Zinn, B. T., *Twenty-fourth Symposium (Intl.) on Combustion*, The Combustion Institute, Pittsburgh, PA, 1992, pp. 1297–1305.
[64] Keller, J. O., Barr, P. K., and Gemmen, R. S., *Combustion and Flame*, Vol. 99, 1994, pp. 29–42.
[65] Keller, J. O. and Hongo, I., *Combustion and Flame*, Vol. 80, 1994, pp. 219.

[66] Logan, P., Lee, J. W., Lee, L. M., Karagozian, A. R. and Smith, O. I., *Combustion and Flame*, Vol. 84, 1991, pp. 93–109.
[67] Smith, O. I., Marchant, R., Willis, J., Lee, L. M., Logan, P. and Karagozian, A. R., *Combustion Science and Technology*, Vol. 74, 1990, pp. 199–210.
[68] Cadou, C., Logan, P., Karagozian, A., Marchant, R., and Smith, O., *Environ. Sensing and Comb. Diagnostics, SPIE Proc. Series* Vol. 1434, 1991, pp. 67–77.
[69] Marchant, R., Hepler, W., Smith, O. I., Willis, J., Cadou, C., Logan, P. and Karagozian, A. R., *Combustion Science and Technology*, Vol. 82, 1992, pp. 1–12.
[70] Willis, J. W., Lee, L.-M., Karagozian, A. R. and Smith, O. I., *Combustion Science and Technology*, Vol. 94, 1993, pp. 469–481.
[71] Willis, J. W., Cadou, C., Mitchell, M., Karagozian, A. R., and Smith, O. I., *Combustion and Flame* Vol. 99, 1994, pp. 280–287.
[72] Pont, G., Willis, J. W., Karagozian, A. R., and Smith, O. I., *Twenty-sixth Symposium (Intl.) on Combustion*, The Combustion Institute, Pittsburgh, PA, pp. 2463–2470, 1996.
[73] Pont, G., Cadou, C. P., Karagozian, A. R., and Smith, O. I., *Combustion and Flame*, Vol. 113, 1998, pp. 249–257.
[74] Cadou, C., Smith, O. I., and Karagozian, A. R., *AIAA Journal*, Vol. 36, No. 9, 1998, pp. 1568–1574.
[75] Kang, Y., Karagozian, A. R., and Smith, O. I., *AIAA Journal*, Vol. 36, No. 9, 1998, pp. 1562–1567.
[76] Zukoski, E. E. and Marble, F. E., *Proc. Gas Dynamics Symp. on Aerothermochemistry*, Northwestern University Press, pp. 205–210, (1956).
[77] Oppelt, T., *Journal of the Air Pollution Control Association*, Vol. 37, 1987, p. 558.
[78] Ho, C. M. and Tai Y.C., *Annual Reviews of Fluid Mechanics*, Vol. 30, 1998, pp. 579–612.
[79] Bowman, C. T., presentation to NAS ASEB Panel on Breakthrough Technologies, January 19, 1998.
[80] K. Bult, *et al.*, "Low Power Systems for Wireless Microsensors", *Proceedings of 1996 International Symposium on Low Power Electronics and Design*, Monterey, CA, Aug. 12–14, 1996.
[81] Hanson, R., presentation to NAS ASEB Panel on Breakthrough Technologies, January 19, 1998.
[82] Sanders, S.T., Wang, J., Jeffries, J.B., and Hanson, R.K., *Applied Optics*, Vol. 40, No. 24, 2001, pp. 4404–4415.
[83] Paschereit, Gutmark, E., and Weisenstein, W., *Combustion Science and Technology*, Vol. 138, No. 1–6, 1998, pp. 213–232.
[84] Cortelezzi, L., Leonard, A., and Doyle, J. C., *J. Fluid Mech.*, Vol. 260, pp. 127–154, 1994.
[85] Cortelezzi, L., *J. Fluid Mech.*, Vol. 327, pp. 303–324, 1996.
[86] Cortelezzi, L. and Speyer, J.L., "Robust Reduced-Order Controller of Laminar Boundary Layer Transitions", *Physical Review E*, Vol. 58, No. 2, 1998, pp. 1906–1910.
[87] Cortelezzi, L., Lee, K.-H., Kim, J. and Speyer, J.L., "Skin-Friction Drag Reduction via Robust Reduced-Order Linear Feedback Control", *International Journal of Fluid Dynamics*, Vol. 11, No. 1–2, 1998, pp. 79–92.
[88] Aubrey, N., Holmes, P., Lumley, J. L., Stone, E., *J. Fluid Mech.*, Vol. 192, pp. 115–173, 1988.

[89] Aubrey, N., Lumley, J. L., Holmes, P., *Theoret. Comput. Fluid Dynamics*, Vol. 1, pp. 229–248, 1990.
[90] Smith, R. S. and Doyle, J. C., *IEEE Trans. Auto. Control*, Vol. 37, No. 7, pp. 942–952, 1992.
[91] Cortelezzi, L., Chen, Y.-C., and Chang, H.-L., *Phys. Fluids*, Vol. 9, No. 7, pp. 2009–2022, 1997.

Nonlinear Modeling and Control of Combustion Dynamics

Anuradha M. Annaswamy

Active-Adaptive Control Laboratory, Department of Mechanical Engineering, MIT Cambridge, MA

Abstract. Active control technology has been used in the context of continuous combustion processes for the past twenty years. Over this time, active control has shown to be successful in reducing unsteady pressure oscillations in a variety of combustion systems ranging from benchtop combustors to large-scale industrial rigs with an energy release of 260 MW. In recent years, it has been shown that it is possible to develop control strategies based on dynamic models that are derived using a combination of physically based principles and input-output measurement data. The models developed include both linear models that capture thermoacoustic instabilities and nonlinear models that characterize stable limit cycles. The approaches for deriving model-based controllers include optimization, adaptation, time-delays, and neural networks. These strategies have been used to modulate fuel-flow and air-flow rates entering the combustion process in a number of practical combustors and shown to result in an order of magnitude improvement in performance. In this article, highlights of the area of active combustion control with particular emphasis on models and model-based control strategies are presented. Both linear and nonlinear models that are reported in the literature are discussed. Linear, time-delay, adaptive, and neural control methods that have been developed thus far are presented. The accruing results and experimental verification in laboratory-scale and large-scale combustors are discussed.

1 Introduction

Continuous combustion processes are prominently encountered in many applications related to power generation, heating and propulsion. Examples are domestic and industrial burners, gas turbines, afterburners, waste incinerators, and scramjet/ramjet engines. Typical requirements in these applications are reduced emissions, large volumetric heat-release, high power density, large turn-down ratios, and high efficiency. As the processes are pushed towards these performance objectives, an ubiquitous characteristic exhibited in several instances is an increased level of pressure oscillations, referred to as thermoacoustic instability. The use of active control where the incoming reactants

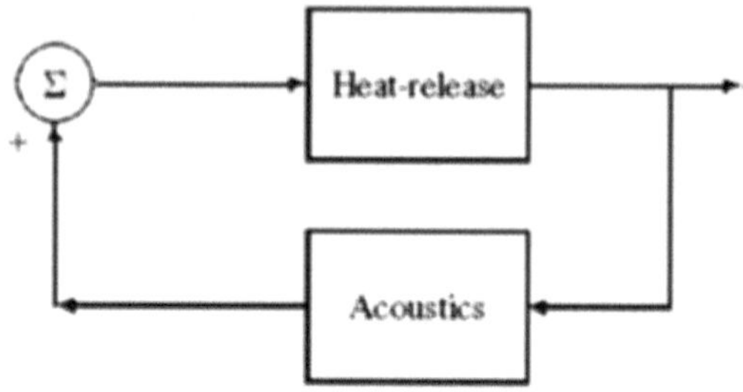

Fig. 1. Thermoacoustic instability in a combustion system

are suitably modulated according to a carefully designed algorithm is becoming increasingly common and attractive for realizing these high performance goals.

A typical continuous combustion process consists reacting mixture flowing in a constant area duct with a flame anchored at a specific location. The latter ignites the reactants, releasing their chemical energy in the form of heat, thus raising their temperature and reducing their density. Flames, which are essentially surfaces across which reactants are converted into products, not only possess their own inherent instabilities, but are also known to respond readily to imposed oscillations. Dynamic instabilities occur in combustion processes due to the feedback coupling between the two dominant processes, acoustics and heat-release (see Fig. 1). The heat release from the combustion of reactants affects the acoustics, and the acoustic perturbations alter the heat release dynamics, closing the loop. As described by Rayleigh in 1845 [1], and later quantified in [2, 3] if the heat-release rate from the combustion process is in phase with the pressure, the system is unstable, and when the heat-release rate is out of phase with the pressure, the system becomes stable. While indeed the above statement pinpoints the basic components in play that induce the dynamic instability, the problem of modeling is a precise quantification, first of the dynamics of these components, and next of all of the coupling mechanisms between these components to produce the stability behavior.

The recent twist in this old problem is that of active control. Simple demonstrations on a Rijke tube in the 80's [4, 5] showed that by using a very small fraction of the system energy, the pressure amplitude can be reduced by several orders of magnitude, thereby establishing the feasibility of active combustion control. Subsequently, several laboratory scale tests were conducted simulating conditions of afterburners, ramjets, and gas turbine dynamics, and attempting active control of the ensuing instability. During 1999–2000, several successes have been reported in large-scale industrial rigs [6, 7, 8], showing that active combustion control is indeed a viable and feasible technology.

Much of the experimental results related to active combustion control have thus far utilized empirical strategies where the idea is to isolate the oscillation frequency, *phase-shift* the signal, and feed it back as a control input so that *noise cancellation* occurs. These methods have shown that active control can lead to a pressure reduction. However, several questions remain regarding the use of such an active control approach, the most important of which is that since this approach is empirical, it is not clear what factors affect the suppression levels the most, or if the achieved levels are indeed optimal for a given type of actuator. The more general question for a combustor of given geometry, mixing conditions, and equivalence ratio, is what the most appropriate actuators are and their most effective locations. A model-based active control strategy enables one to analyze the behavior of the combustor, quantify its characteristics in terms of parameters such as geometry, equivalence ratio, flow rate, flame stabilization mechanism, and heat loss, thus enabling one to optimize the control design against the desirable objectives.

Carrying out a thorough analytical investigation of the behavior of a combustion process is an exceedingly difficult task since the fundamental principles governing its dynamics have to be brought together from a number of disciplines including acoustics, combustion dynamics, hydrodynamics, thermodynamics, and heat transfer. In addition, the investigation must include a study of how these distributed systems behave in the presence of localized control sources which introduce time-varying boundary conditions in the process. The distributed nature of these processes makes partial differential equations that are highly nonlinear and coupled as their governing entities. For the purpose of understanding the dominant dynamic behavior of the system as well as designing active control inputs, reduced-order models are most useful. In recent years, significant progress has been made towards modeling the dominant dynamics that are responsible for producing the thermoacoustic instability in combustion processes.

In this paper, highlights of modeling and model-based control of combustion dynamics are presented. While details of linear modeling theory and experimental results that have been reported in the literature can be found in [9], in this article, our emphasis is on nonlinear models and control, and in particular, on models that have been explicitly used for control design and experimental implementation. Experimental results that have been reported in the literature using these model-based controllers are also described in this paper. Section 2 addresses models of combustion dynamics while Sect. 3 deals with model-based control methods. In Sect. 4, summary and concluding remarks are presented.

2 Dynamic Models of Combustion Instability

Dynamic instabilities occur in combustion processes due to feedback coupling between two dominant processes, acoustics and heat-release (see Fig. 1). Significant research has been carried out in modeling the detailed dynamics of these two processes and their feedback interactions, and are described below.

2.1 Linear Models

Acoustics represents the host oscillator, that leads to large and sustained unsteady pressure oscillations. In order to model the underlying acoustics, utilizing the predominantly one-dimensional configurations typically present in combustors, the acoustic equations for longitudinal, bulk, and azimuthal modes can be derived using the standard wave equation [10, 11]. Using an expansion

$$p'(x,t) = \overline{p} \sum_{i=0}^{n} \psi_i(x)\, \eta_i(t), \tag{1}$$

where ψ_i denote basis functions of the form $\psi_i(x) = \sin(k_i x + \phi_{i0})$, $i = 1, \ldots, n$, and k_i and ϕ_{i0} determined from the boundary conditions, and if the heat-release rate is the only dominant source that is localized at x_f, the modal amplitudes can be shown to satisfy the following equation [12]:

$$\ddot{\eta}_i + 2\zeta\omega_i\, \dot{\eta}_i + \omega_i^2 \eta_i = \sum_{i=1}^{n} \widetilde{b}_i \dot{Q}' \tag{2}$$

where Q is the total heat-release rate, $\widetilde{b}_0 = \gamma - 1$, $\widetilde{b}_i = \gamma V a_o \psi_i(x_f)/E$ for $i = 1, \ldots, n$, $E = \int_0^L \psi_i^2(x)\, dx$, γ is the specific ratio, $a_o = \frac{\gamma - 1}{\gamma \overline{p}}$, ζ represents the passive damping ratio in the combustor[1], L is its length, $\omega_0^2 = \sqrt{\frac{A_n}{L_n V}}$, V is the volume of the combustor, A_n and L_n are the cross-sectional area and length of the inlet/outlet neck connected to the combustor. and $\omega_i = k_i \overline{c}$, $i = 1, \ldots, n$. Typically, $\omega_0 << \omega_i$, for $i = 1, \ldots, n$. The acoustics of azimuthal modes can be computed in a similar manner. For acoustic modes in a dump combustor, which is a commonly used configuration, the spatial modes can be determined by considering the cold and hot sections separately, and are not necessarily in the form of a sinusoid. It should be noted that in the case of a dump combustor, the corresponding equations will assume a more complex form since the underlying basis functions are not necessarily orthogonal. An alternate procedure for deriving the acoustics equation is to adopt a wave approach, as in [13], which has been used successfully to design model-based controllers [14].

The second process that contributes to the combustion dynamics is due to unsteady heat release. Models of unsteady heat-release have been derived at two different conditions, which correspond to high and low Damkholer numbers, respectively. At high Damkohler numbers and weak to moderate turbulence intensity, the heat release can be represented as a thin flame. The flame surface is described by a single-valued function $\xi(r,t)$ which represents

[1]Dissipation in a combustor can be caused by heat losses in the flame zone and friction due to viscous effects.

the instantaneous axial displacement of the flame, [2], and the total heat release rate, Q, is proportional to the integral of this surface over an anchoring ring:

$$Q = \kappa(\phi) A_f(r,t) \tag{3}$$

$$A_f = 2\pi \int_0^R r\sqrt{1+\left(\frac{\partial \xi}{\partial r}\right)^2} dr, \qquad \kappa(\phi) = \rho_u S_u(\phi) \Delta h_r(\phi) \tag{4}$$

$$\frac{\partial \xi}{\partial t} = u - v\frac{\partial \xi}{\partial r} - S_u(\phi)\sqrt{\left(\frac{\partial \xi}{\partial r}\right)^2 + 1}, \tag{5}$$

where u and v are velocities in the x and y directions, ϕ is the equivalence ratio, S_u is the burning velocity, ρ_u is the density of the unburnt mixture, Δh_r is the heat of reaction, and A_f denotes the flame area.

Since unsteady heat-release can be produced in response to both perturbations u' and ϕ' in the flow velocity and equivalence ratio, respectively, the linearization of (3)–(5) with respect to both quantities must be considered, which leads to a linear model of the form [15]

$$\dot{Q}' = d_0 u' + d_1 \left(u'_{\tau_f}(t)\right) + d_2 \left(\phi'_{\tau_f}(t)\right) + d_3 \phi' + d_\phi \dot{\phi}' \tag{6}$$

where

$$x_\tau(t) \triangleq \int_{t-\tau}^{t} x(\zeta) d\zeta, \tag{7}$$

$$d_0 = \bar{\kappa} R, \quad d_1 = -\bar{\kappa}\, \bar{S}_u, n \quad d_2 = -\bar{\kappa} \bar{S}_u \left.\frac{\partial S_u}{\partial \phi}\right|_{\bar{\phi}} \left.\frac{\partial \xi(r)}{\partial r}\right|_0,$$

$$d_3 = -\bar{\kappa} \frac{\partial \bar{S}_u}{\partial \phi} \bar{\xi}(0), \quad \bar{\kappa} = 2\pi\rho \Delta h_r \bar{S}_u, \quad \tau_f = \frac{R}{\bar{S}_u},$$

$$d_\phi = 2\pi\rho_u \left(\bar{S}_u \left.\frac{d\Delta h_r}{d\phi}\right|_{\bar{\phi}} + \Delta \bar{h}_r \left.\frac{dS_u}{d\phi}\right|_{\bar{\phi}} \right) \left(\int_0^R r\bar{\xi} dr \right).$$

τ_f represents the characteristic propagation delay of the flame surface into the reactants flow. If only u' perturbations are significant, it can be shown that the unsteady heat-release is given by [16, 12]

$$\dot{A}'_f = -2\pi S_u \int_{t-\tau_f}^{t} u(t')dt' + 2\pi R u' \tag{8}$$

$$\approx -b_f A'_f + 2\pi R u' \qquad b_f = 2\frac{S_u}{R}$$

$$Q' = \bar{\kappa} A'_f$$

[2]We consider here a flame stabilized over a perforated plate, R is the radius of the perforation.

At large turbulence intensities, assuming that the combustion zone can be modeled as a Well Stirred Reactor [17], with the reactant and product thoroughly mixed and indistinguishable in the combustion volume [18, 19], the response to velocity perturbations can be modeled as [20]

$$\dot{Q}' + \alpha Q' = \beta u' \tag{9}$$

where α and β are related to the time-constant of the evolution and consumption of mass flow rates of various species.

Perturbations in equivalence ratio could be introduced due to the feedline dynamics in the following manner. For a choked fuel-nozzle and an unchoked air-nozzle, the instantaneous equivalence ratio ϕ_s at the exit of the fuel nozzle due to air flow fluctuations is determined as

$$\phi_s = \frac{\overline{\phi}}{1 + u_s'/\overline{u}} \tag{10}$$

where u_s denotes the velocity at the fuel supply, and is similar to a relation used in [21]. When linearized, we obtain the relation [21]

$$\phi'(t) = -\frac{\overline{\phi}}{\overline{u}} u_s'(t - \tau_c) \tag{11}$$

where ϕ' denotes the equivalence ratio perturbation at the burning zone, τ_c is a convective delay due to transport lag from the supply to the burning plane of the flame, and is given by $\tau_c = L/\overline{u}$, where L is the distance from the supply to the burning plane. The coupling between u and p can be determined using the conservation equations

$$\frac{\partial p'}{\partial t} + \gamma \overline{p} \frac{\partial u'}{\partial x} = (\gamma - 1) q' \tag{12}$$

$$\frac{\partial u'}{\partial t} + \frac{1}{\rho} \frac{\partial p}{\partial x} = 0\,. \tag{13}$$

With the acoustics, heat-release, and coupling effects modeled as above, we obtain the following model of combustion instability:

$$\ddot{\eta}_i + 2\zeta\omega_i\dot{\eta}_i + \omega_i^2\eta_i = \frac{\widetilde{b}_i}{V}\Big[d_0 u' + d_1\left(u'_{\tau_f}(t)\right) + d_2\left(\phi'_{\tau_f}(t)\right) + d_3\phi'(t) + d_\phi\dot{\phi}'(t)\Big] \tag{14}$$

$$\phi'(t) = \sum_{i=1}^{n} d_i \eta_i (t - \tau_c) \tag{15}$$

$$u' = \sum_{i=1}^{n} \widetilde{c}_i \dot{\eta}_i + \theta a_0 Q' \tag{16}$$

where

$$\widetilde{c}_i = \frac{d\psi}{dx}(x_f)\frac{1}{\gamma k_i^2}, \qquad d_i = \frac{d\psi}{dx}(x_s)\frac{\overline{\phi}}{\rho\overline{u}},$$

and θ represents the effect of the flow velocity before and ahead of the flame. Equation (14) indicates that two different coupling mechanisms are possible excitations for the acoustics, one resulting from the velocity perturbations u' and the other from equivalence ratio perturbations ϕ'. Equation (14) also indicates that two different time-delays, τ_f and τ_c can induce these excitations, one arising from propagation effects, and the other from convection. An additional point to note from (14) is that if the dominant pressure mode is that of a bulk-mode, then it can be excited only due to perturbations in the equivalence ratio. However, if longitudinal modes are the ones that are dominant, they can be excited either by u'-perturbations or by ϕ'-perturbations.

The structure of these equations show that the combustion dynamics in general is given by a $2n+1$th order ordinary differential equation with time-delays τ_f and τ_c. In the simplest case, when only one acoustic mode is dominant, setting $\eta_i = \eta$, the combustion dynamics can be shown to be of the form

$$\ddot{\eta} + (2\zeta\omega - \gamma_1)\dot{\eta} + (\omega^2 + \gamma_2)\eta - \gamma_2\eta(t - \tau_f) = 0 \tag{17}$$

if the variations are mainly in u', where

$$\gamma_1 = 2\pi\rho R\Delta h_r \widetilde{b}\widetilde{c}, \qquad \gamma_2 = 2\pi\rho R\Delta h_r \overline{S}_u \widetilde{b}\widetilde{c}$$
$$\widetilde{c} = \frac{1}{\gamma k^2}\frac{d\psi}{dx}\overline{p}, \qquad \widetilde{b} = \gamma a_o \frac{\psi(x_f)}{E}, \qquad \tau_f = \frac{R}{S_u}$$

and of the form

$$\ddot{\eta} + 2\zeta\omega\dot{\eta} + \omega^2\eta - \beta_1\eta(t - \tau_c) = 0 \tag{18}$$

if they are due to ϕ', where

$$\beta_1 = 2\pi\widetilde{b}\frac{\overline{\phi}}{\overline{u}}\widetilde{c}\gamma k^2 \left(\overline{S}_u \left.\frac{d\Delta h_r}{d\phi}\right|_{\overline{\phi}} + \Delta\overline{h}_r \left.\frac{dS_u}{d\phi}\right|_{\overline{\phi}} \right) \left(\int_0^R \xi dr \right), \qquad \tau_c = \frac{L}{\overline{u}}$$

Both (17) and (18) have been shown to predict instability, match experimentally observed results [12, 15, 22], shown to explain anomalies such as secondary peaks with empirically designed controllers [23], and lead to optimal model-based controllers in experimental investigations [24, 25].

2.2 Nonlinear Models of Combustion Oscillations

Nonlinear features are abundant in a combustion process. The most dominant of these is a limit-cycle behavior which is exhibited by almost all the variables in the process including pressure, velocity, and heat-release. The typical dynamic response of any of these variables consists of a divergent set of oscillations which transition then to a sustained periodic signal, that is almost sinusoidal in nature. Several speculations have been made regarding mechanisms responsible for such a behavior. Nonlinearities in the heat-release dynamics have been noted in [13, 21, 26, 27, 28] while nonlinearities in acoustics are claimed to be responsible for these limit-cycles in [11, 29, 30]. Below, we discuss these further and present plausible nonlinear models that describe some of these mechanisms.

The presence of such limit-cycles, many of whom are stable, suggests the obvious presence of bifurcations. One of the key parameters that appears to induce these bifurcations is the mean equivalence ratio, ϕ. Two distinct ranges of ϕ appear to be of interest, depending on the application. In ramjet engines and afterburners, close to stoichiometry, instability appears to result followed by the presence of supercritical bifurcation. In engines with strict emission requirements, as one attempts to burn lean, a "blow-out" limit is reached which once again is accompanied by these bifurcations. In many of these cases, more than one limit-cycle is encountered [28] suggesting the presence of both sub- and supercritical bifurcations.

Finally, in [28, 31, 32, 33], hysteresis mechanisms have been observed and discussed. The parameters in question are the mean equivalence ratio and the mean inlet velocity. In [32], keeping other parameters a constant, as ϕ is increased steadily and then decreased, the behavior at the same value changes from instability to stability which is sometimes accompanied also by a drastic change in the flame structure. In [32] and [33], it is shown that once such a mechanism is present, appropriate use of it can be made in designing active control strategies and reduce the amplitude of oscillations.

Of all the nonlinear mechanisms in combustion processes, it is limit-cycles that have been examined the most. As mentioned earlier, nonlinearities in both acoustics and heat-release have been analyzed and shown to result in limit-cycles. Since in most of the combustion systems, the pressure amplitudes are in general small, nonlinearities due to the acoustics can be assumed to be negligible [34]. We consider in this paper, the latter case, and begin with the linear acoustics model as in (2) and the nonlinear heat-release model given by (3)–(5) and the coupling by the relation (10). While different sources contribute to the heat-release nonlinearity, at least three different mechanisms have been studied and analyzed in the literature. The first two concern nonlinearities in the area as well as the burning velocity of the flame, resulting in a nonlinear response of the heat-release rate as the incoming velocity increases [13, 26]. The third is due to the varying fuel-air ratio that is delivered to the flame [21] resulting from the feedline dynamics. We briefly state each of these nonlinearities below:

1. Nonlinear Burning Velocity: The starting point for the model in [26] is (5) which is reduced to a nonlinear ordinary differential equation by using an expansion

$$\xi^* (r,t) = \overline{\xi} + R \sum_{j=1}^{m} x_{q_j} (t) \psi_{f_j} (r) , \tag{19}$$

$$\psi_{f_j} (r) = \left(\frac{r}{R}\right)^{j-1} , j = 1, 2, \ldots, m . \tag{20}$$

The choice of $\psi_f(\cdot)$ is made based on flame structures typically found in weakly turbulent flames. When substituted in (5), (19)–(20) lead to the differential equation

$$\sum_{j=1}^{m} B_{jk} \dot{x}_{q_j} = -A_k + U_k \tag{21}$$

$$A_k = \int_0^R \psi_{f_k} S_u A_r(x) dr, \quad A_r(x) = r \sqrt{1 + \left(\frac{d\overline{\xi}}{dr} + R \sum_{j=1}^{m} x_{q_j} \frac{d\psi_{f_j}}{dr} \right)^2}$$

$$U_k = \int_0^R r \psi_{f_k} u' dr, \qquad B_{jk} = R \int_0^R r \psi_{f_k} \psi_{f_j} dr$$

$$Q = c_0 \int_0^R A_r(x) dr, \qquad c_0 = 2\pi \rho_u \Delta h_r \tag{22}$$

The quantities that were unspecified in the flame surface equation are $\overline{u}$, $\overline{\xi}$, and S_u which are in general spatially distributed. The exact nature of the functions is exceedingly difficult to determine since they depend on heat losses, flame stretch and flow entrainment. Using the flow field profile of a laminar jet [35], we have

$$\overline{u}(r) = k_0 \frac{1}{(k_1 + \overline{\xi}) \left(1 + \frac{\overline{\xi}^2}{4}\right)}, \qquad \overline{\xi}(r) = k_2 r \tag{23}$$

where $k_i, i = 0, 1, 2$ are suitably chosen constants. Also, since the local burning rate depends on the heat loss to the stabilizing disc as well as the oscillating flow, S_u depends on both the location of the flame and the velocity perturbation, and is chosen as

$$S_u \left(\overline{\xi}, u'\right) = \overline{S}_u \left(1 - e^{c_1 \overline{\xi}}\right) + c_2 |u'|^3 u' \tag{24}$$

where c_1 and c_2 are suitably chosen positive constants. The exponential term emphasizes the exponential dependence of the reaction rate on the temperature, while the cubic nonlinearity stems from the dependence of the heat transfer coefficient on the flow velocity.

It can be shown that for suitable choices of the parameters, the heat-release model as in (21) together with the linear acoustic model in (14) leads

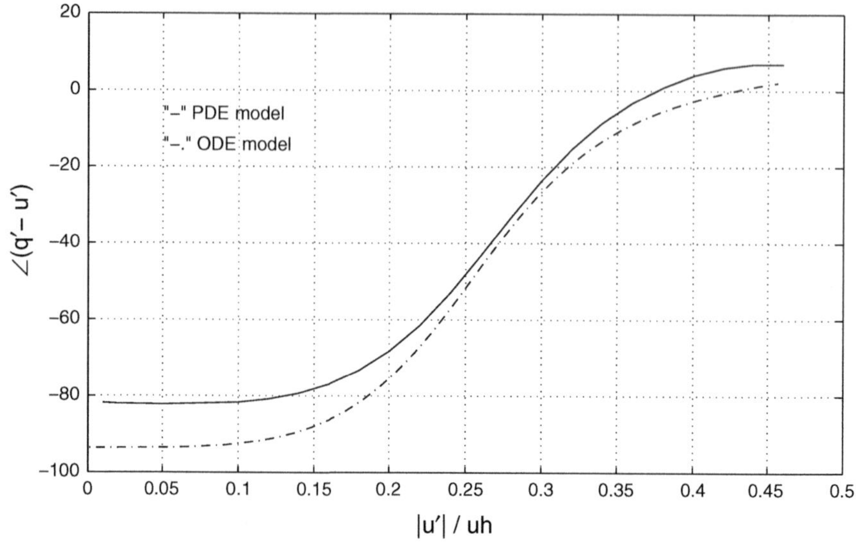

Fig. 2. Change in the phase between the velocity u' and the heat-release rate Q' as the amplitude $|u'|$ increases in the presence of a nonlinear S_u

to limit cycles when only u'-perturbations are present. It is also interesting to note that the relation between u' and Q' is characterized by phase rather than gain; as the amplitude of u' is increased, the phase between Q' and u' was observed to change from –90° to 0° (see Fig. 2). It was observed that the limit-cycle behavior persisted as $\overline{\phi}$ and the flow-rate was changed in the model over a large range of values, with the limit-cycle amplitude increasing as ϕ increased. Preliminary studies from an experimental investigation of a 1kW combustor support these phase-changes [36].

2. Saturation nonlinearities: Yet another nonlinearity that affects the heat-release rate is saturation, and was suggested by Dowling in [13]. which appears in the following way: While for small amplitudes, $Q(t)$ is linearly proportional to $u(t)$, as the amplitudes increase, the flow-velocity reverses and becomes negative. However, the physical constraint on the heat-release dictates that Q remain positive, which implies that $Q(t)$ saturates. Such a flow-reversal is seen occur in the experimental rig containing the bluff-body stabilization [37]. The underlying model can be expressed in this case as

$$\dot{Q} = -b_f Q + f_2(u') \tag{25}$$

where

$$\begin{aligned} f_2(u) &= u \ \text{ if } \ |u| \leq u_0 \\ &= u_0 \text{ if } \ |u| > u_0 \end{aligned} \tag{26}$$

An analysis of the acoustics as in (2) together with (25) and coupling as in (16), with the equivalence ratio fluctuations assumed to be small (due to proper mixing, for example), can be shown to lead to limit cycles [13, 27]. In [13], a comparison is made with experimental results of [37], and the limit-cycle characteristics are shown to be similar.

3. Fuel-air ratio nonlinearities: In [21], starting from the flame kinematic model as in (3)–(5), on the basis of experimental observations, the area nonlinearities are neglected, while the change in the heat of reaction due to changes in ϕ as well as the nonlinearity due to the fuel-air delivery as in (10) are retained. Assuming that only one acoustic mode is dominant, this yields a representation of the form

$$
\begin{aligned}
\ddot{\eta} + \omega^2 \eta &= b \frac{d}{dt} \left[\left(\overline{A}_f + A'_f \right) f_3(\phi(t)) \right] \\
\dot{A}'_f &= -b_f A'_f + 2\pi R u' \\
f_3(\phi) &= k_1 (\phi - k_2)^p \exp\left(-(1 - \phi + k_3)^2 \right) \\
\phi &= \frac{\overline{\phi}}{1 + u'(t - \tau_c)/\overline{u}} \qquad u'(t) = k_4 \dot{\eta}
\end{aligned}
\tag{27}
$$

where constants k_i represent a variety of physical features including the change in the heat of reaction as ϕ approaches the blow-out limit and acoustic admittance, and τ_c is a convective time-lag. It can be shown that for small amplitudes, the above system exhibits instability for banded values of τ_c [31], and for larger amplitudes, it exhibits limit cycles. A comparison of the model with a single nozzle rig over a range of pressures, temperatures, and equivalence ratios is made in [21], where it is shown that the limit-cycle amplitudes match the model prediction fairly well. As ϕ increases, a 20% mismatch is shown in the frequency prediction, while the time-delay prediction is fairly accurate at low equivalence ratios.

2.3 Linear Models Using a System Identification Approach

Even though the combustion response is clearly nonlinear, in an experimental run, one seldom captures the signal growth within the linear range and transition phase due to its brevity. It is the periodic pressure/heatflux signal, which is the more persistent feature and the one that is experimentally recorded. If it is the periodic oscillations that need to be modeled, one can choose a linear model structure and a system identification procedure to identify the model parameters that capture the pressure characteristics. In [38]–[41] and [25], such an approach is adopted. Assuming that the control input that corresponds to an actuator output is denoted as V and the resulting pressure by p', the system identification model structure can be chosen as

$$
\dot{x} = Ax + bV + dW \qquad p' = c^T x \tag{28}
$$

where A, b, c are model parameters that are to be identified using input-output data, x is the state of the system, and W is an exogenous noise signal. The dimension of x is to be chosen *a priori* so that the model order is commensurate with the dominant dynamics that needs to be modeled. By using a pseudo-random-binary signal in V, these parameters can be identified accurately.

It should be pointed out that the linear model has been derived under the following assumptions: i) the limit cycle response can be approximated by the homogeneous response of a linear differential equation; ii) the effect of an external input on the pressure response can be represented by a linear term. Since (a) the nonlinear response in combustion systems is a stable limit cycle, (b) the sustained pressure response, in most cases, is sinusoidal in nature, and (c) since the effect of a small external input, in general, is small compared to the unforced response, both assumptions (i) and (ii) are reasonable for pressure responses that are close to the limit cycle. This implies that any model-based controller is guaranteed to be accurate in a neighborhood of the limit cycle, and therefore can reduce its size. The extent of this neighborhood depends on how well the gain and phase of the linear model can approximate that of the nonlinear system as the size of the limit cycle reduces. For all amplitude levels of the pressure where this approximation error remains small, the model-based based controller will continue to reduce the amplitude further until the pressure amplitude reaches background noise levels. In [25], it has been shown that such an approach can be used successfully to suppress pressure oscillations down to background noise in a 30 kW swirl-stabilized combustor.

3 Control of Combustion Instability

The most commonly used controls strategy to reduce combustion oscillations is a phase-shift controller that consists of isolating the pressure component at the unstable frequency through proper filtering, adding a phase-shift, and feeding the signal back to modulate a suitably chosen control input. Typical inputs are secondary fuel and secondary air that are pulsed using fuel-injectors and acoustic drivers, respectively. While these controllers are quite successful [6, 42], quite often their scope of performance is constrained in terms of level of improvement and range of operating conditions. At some of the operating conditions, secondary peaks are generated due to the control action thereby compromising on the maximum damping that is achievable. If more than one frequency is dominantly present, the control design seems to prove quite challenging. In some cases, the phase-shift controller appears to be quite sensitive to perturbations. Model-based controllers are currently being implemented with success over a range of rigs, and are shown to result in an order of magnitude improvement in the performance. The reader is referred to [9] for further details.

In this section, we focus on the highlights of these model-based controllers, and show how each aspect of the model is accommodated in the control design. In particular, we discuss the behavior of (i) a linear controller in the presence of stable limit-cycles, (ii) a linear time-delay controller that provides stabilization even in the presence of large time-delays, (iii) an adaptive time-delay controller that combines the properties of the controller in (ii) and accommodates parametric uncertainty, and (iv) a neural controller that accounts for uncertainties in the nonlinearity.

The first step in the design of a model-based controller is the quantification of the effect of the active control input on the combustion dynamics. Examples of typical active control devices are fuel-injectors and loudspeakers, where the former has a dominant effect of additional mass-flow which results in additional heat-release, and the latter introduces additional velocity which impacts on both acoustics and heat-release. If the quantity added is fuel, in addition to the mass-flow, heat input is introduced as well, since it changes the equivalence ratio. Defining

$$\phi_c = \frac{\dot{m}'_c}{\overline{\dot{m}}_a \phi_0}$$

where $\dot{m}'_c$ is the perturbation in the fuel-flow rate, $\overline{\dot{m}}_a$ is the mean air mass flow rate, and ϕ_0 is the fuel to air ratio at stoichiometry, and v_c as the velocity of the diaphragm of the loudspeaker, and neglecting the time-delays τ_c and τ_f in the combustion model, the effect of the active control input on the combustion dynamics can be quantified as follows: If a fuel-injector is used as an actuator, it can be shown that when only ϕ'-perturbations are present, in the linear case, the heat release dynamics in (6) is altered as

$$\dot{Q}' = d_2 \left(\phi'_{\tau_f}(t) + \phi'_{c_{\tau_f}}(t - \tau_{co})\right) + d_3(\phi' + \phi_c(t - \tau_{co})) + d_\phi(\dot{\phi}' + \dot{\phi}_c(t - \tau_{co}))$$

where $\tau_{co} = L_c/\overline{u}$, and L_c is the distance between the burning plane and the location of the fuel-injector. For the case when a speaker is used as the actuator, assuming that no equivalence ratio perturbations are present, it can be shown that the above equations simplify to the form [24]

$$\ddot{\eta}_i + 2\zeta_0\omega\eta_i + \omega_i^2\eta_i = b_i\, \dot{q}'_f + b_{c_i}\dot{v}_c \tag{29}$$

$$y = \sum_{i=1}^{n} c_{c_i}\eta_i, \tag{30}$$

$$\dot{q}'_f = \sum_{i=1}^{n} \left(f(g_f c_i\, \dot{\eta}_i + k_{ao}\alpha_r v_c)\right)$$

In addition to the above model, the actuator dynamics itself must be included in the control design. In the case of a fuel-injector, the mechanical and fluid parts of the device can be modeled as [15]

$$\frac{\dot{m}'_c(s)}{V(s)} = \frac{k_m}{\tau_m s + 1}, \tag{31}$$

where V is a voltage applied to the injector, while in the case of a speaker, the transfer function between the voltage applied and the acceleration $\dot{v}_c$ can be derived as [24]

$$G_l(s) = \frac{k_1 s^2}{s^2 + 2\zeta_l \omega_l s + \omega_l^2} \tag{32}$$

where k_l, b_l, and m_l denote the stiffness, friction, and mass properties, and k_1 is a calibrating gain.

The discussions in Sect. 2 show that combustion dynamics predominantly exhibit two characteristics, with the first being linear instability which results in diverging pressure oscillations which are then followed by a nonlinear limit-cycle which results in the oscillations becoming sustained. In Sect. 2.1, it was shown that the instabilities could be produced either due to a time-delay or due to negative damping which is produced due to the phase-lag between the unsteady heat-release rate and the unsteady velocity. In Sect. 2.2, we noted that limit-cycles can be produced by more than one type of nonlinear mechanism that may be present in the heat-release response. While both the saturation and fuel-air nonlinearities produce gain-changing mechanisms, the nonlinearity in the burning velocity seems produce a phase-changing nonlinearity. In different rigs, as well as under different operating conditions in the same rig, it is quite possible that the actual mechanism that produces the limit cycle may vary. In practice, therefore, the nonlinear phenomenon responsible for the limit-cycle behavior may not be possible to identify or vary with operating conditions. This implies that if a controller design relies on the structure of f, then the accuracy of the controller and the resulting closed-loop performance can be directly compromised by the lack of fidelity in the model. In such cases, it may be more advantageous to use a linear controller which is designed by entirely neglecting the nonlinearity and using the linear model only.

In order to evaluate the behavior of the closed-loop system with the nonlinear model together with a linear controller, we focus our attention on the case when only u'-perturbations are present, neglect the time-delays, which leads to a feedback system of the form

$$\begin{aligned} \widetilde{u}'_f &= G(s)[\widetilde{q}' + V] \\ \widetilde{q}' &= -f(\widetilde{u}'_f) \end{aligned} \tag{33}$$

where $G(s)$ represents linear dynamics in the acoustics and heat-release dynamics as well as in the actuator, and f represents the nonlinearity.

3.1 Linear Control

Suppose a linear controller is designed using a linearized model of (33) as

$$V = G_c(s)y \tag{34}$$

to achieve closed-loop stability. Expressing the nonlinear function f as

$$f(u) = u - g(u),$$

the resulting closed-loop system can be described as (see Fig. 3)

$$y = W_{cl}(s)\left[-g\left(y\right)\right] \tag{35}$$

where $W_{cl}(s)$ represents the stabilized closed-loop system with the linear controller, whereas $g(\cdot)$ represents the deviation in f from linearity. Under certain conditions on $W_{cl}(s)$ and $g(\cdot)$, it can be shown that the closed-loop system will be stable. The following theorem summarizes these conditions [43]:

Theorem 1. *If the components in the closed-loop system in Fig. 3 are such that*

$$\begin{array}{lll} (i) & W_{cl}(s) & \text{is strictly positive real,} \\ (ii) & g(x)\,x > 0 & \text{for all } x \neq 0 \\ & g(0) = 0 & \end{array}$$

Then the closed-loop system is globally asymptotically stable.

3.1.1 Experimental Results

We present two different experimental demonstrations of a linear LQG-LTR controller [44], where in (1), the controller is designed using the model in (29)–(30) and implemented in a 1 kW combustor with a constant area of cross-section, while in (2), the controller is designed using a system-identification model as in (28) and implemented in a 30 kW swirl-stabilized combustor.

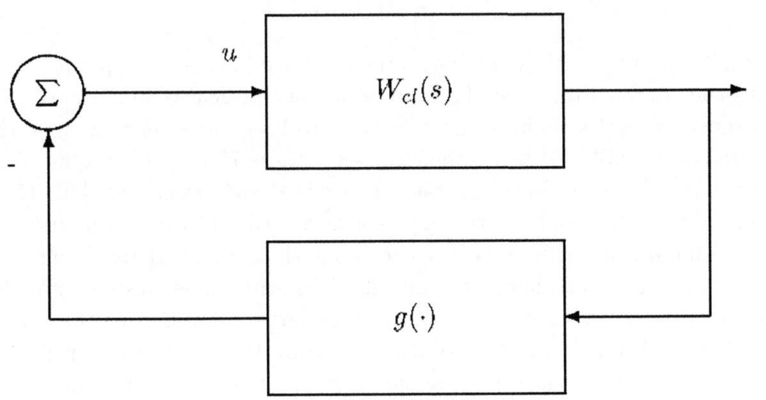

Fig. 3. Stable nonlinear feedback systems

The results in both cases show that the nonlinear combustion process can be successfully controlled using a linear controller.

(1) A bench-top tube-combustor rig was constructed to evaluate the model-based approach to control design. A condensor microphone was used as the sensor and a 0.2W loudspeaker was used as an actuator. Measurements on the test rig were recorded using a Keithley MetraByte DAS-1801AO data acquisition and control board, with a maximum sampling frequency of 300 KHz. Most experiments were conducted with an equivalence ratio between 0.69 and 0.74 and an air flow rate of 333 mL/s (0.38 g/s), which corresponded to an unstable operating condition without control. The pressure response showed a limit-cycle behavior at all of these operating conditions. Linear LQG-LTR and H_∞ strategies were implemented to suppress the pressure oscillations. A 50 dB pressure reduction was achieved using the former using a peak electrical power of 3 mW (see [24] for further details). These results show that a linear model-based controller is successful in suppressing the pressure oscillations despite the presence of the nonlinearities.

(2) In [45], experiments were performed in a swirl-stabilized combustor operating at 30 kW heat release to validate model-based controllers. The average fuel flow rate was kept constant at 0.75 mL/sec. Primary air with a flow rate of 0.056-0.283 m3/sec, at five atmospheres, was used to atomize the fuel. The fuel stream was modulated using a Bosch fuel injector driven by a signal processor, and pressure and heat-release measurements were made. The combustor exhibited a stable limit-cycle at almost all of the operating conditions. A linear-model based LQG-LTR controller was designed and implemented, which brought the pressure oscillations down to background noise.

3.2 Time-delay Control

In the presence of an actuated input ϕ_c that may be due to a fuel injector, the combustor model in (18) gets modified as

$$\ddot{\eta} + 2\zeta\omega\dot{\eta} + \omega^2\eta - \beta_1\eta(t - \tau_c) = k\phi_c(t - \tau_{co}) \tag{36}$$

where τ_{co} is the transport delay due to the distance between the point of injection and the combustion zone. The model in (36) indicates that the requisite control must explicitly include time-delays into its control design, since these delays are comparable to the system time-constants if not larger. One control strategy that can be employed in this context is Posi-Cast control [46, 47, 48]. The idea behind this control strategy is forecast the future output using the system model and use this in turn to stabilize the system. It has been shown that (i) Posi-Cast controllers are successful in stabilizing system models of the form of (36), even when τ_c is significantly large compared to the acoustic time-constants [15], (ii) stability robustness properties of these controllers can be quantified [49], (iii) adaptive posi-cast controllers can be derived to accommodate parametric uncertainties [14, 49]. The highlights of this approach are summarized below.

The plant to be controlled is assumed to be of the form

$$y(t) = W_p(s)u(t - \tau) \tag{37}$$

with the goal of reducing the plant output to zero. In such a case, the following controller can be shown to be stabilizing [15]:

$$V(t) = \frac{c(s)}{\Lambda(s)}V(t - \tau) + \frac{d(s)}{\Lambda(s)}p'(t) + V_1(t), \tag{38}$$

$$V_1(t) = \sum_{i=1}^{n}\left(\int_{-\tau_c}^{0} e^{-\lambda_i\sigma}V(t + \sigma)d\sigma\right), \tag{39}$$

where λ_i's are the poles of $W_p(s)$, $\Lambda(s)$ is a chosen stable polynomial of degree $n-1$, $d(s), n_1(s)$ and $n_2(s)$, are polynomials of degree $n-1$ at most, and $c(s)$ is of degree $n-2$. It can then be shown that for suitable values of the polynomials c, d, n_1, and n_2, the plant can be stabilized for any time-delay τ [48].

3.2.1 Simulation Results

The controller in (38) and (39) was implemented assuming that the fuel-injector is located at a distance of ~ 3 cm upstream the burning zone. τ_c is estimated to be 100 ms, which is about 50 times the time constant of the unstable frequency. The closed-loop simulation is illustrated in Fig. 4. Although control is switched *on* at $t = 50$ ms, the pressure keeps increasing for an additional $t = \tau_c = 100$ ms (from $t = 50 - 150$ ms), then stalls for another 100 ms

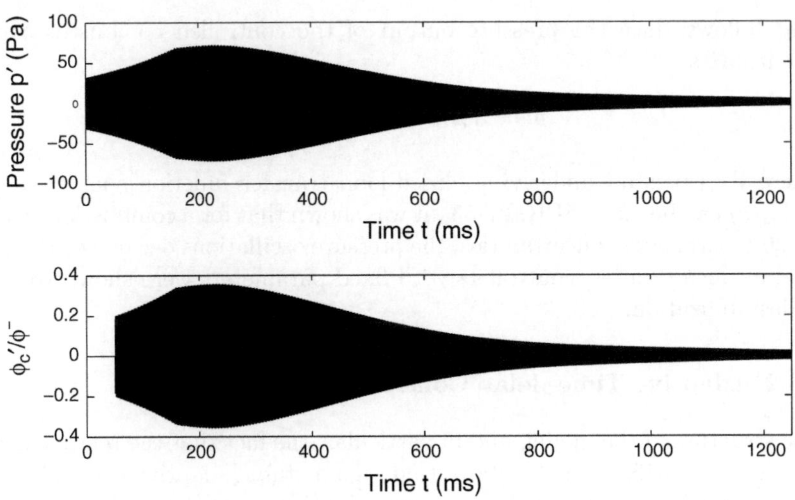

Fig. 4. Response of the controlled combustor with a time-delay of 100 ms in the input signal, proportional injector

(from $t = 150 - 250\,\mathrm{ms}$) before decaying. The reason for the former delay is physical and is due to the time taken for the pulsed-fuel to reach the burning zone. The latter is due to a computational delay in the controller.

3.3 Adaptive Control

Active adaptive control of combustion using the well known LMS algorithm [50] has been implemented in the literature [51, 52] with mixed results. In both cases, the measured output was the acoustic pressure and the heat release, and the actuator was an acoustic driver which was used to modulate either the fuel or the airflow rate. In [51, 52], it was observed that while in some cases, this adaptive approach was successful, in others, the adaptive filter led to divergence in the filter coefficients. In contrast to the above, model-based self-tuning control methods provide guaranteed stability and lead to improved performance, and is discussed below.

3.3.1 Model-based Self-tuning Control

For a plant of the form $y = W_p(s)u$, if the plant is minimum phase, and is of relative degree two, the following control strategy guarantees stabilization:

$$u = k_0(t)\left[\overline{p}'(t) + \dot{k}_0\overline{p}``(t)\right]$$

$$\overline{p}' = k_c\frac{s+z_c}{s+p_c}[p'], \qquad \overline{p}`` = \frac{1}{s+a}\overline{p}'$$

$$\dot{\widetilde{k}}_0 = -\gamma_k p'\overline{p}`` \qquad \gamma_k > 0$$

This follows since the pressure output of the controlled combustor can be rewritten as

$$p' = W_{cl}(s)(s+a)\left[\widetilde{k}_0\overline{p}``(t)\right]$$

where $W_{cl}(s)$ is an underlying closed-loop transfer function such that $(s+a)W_{cl}(s)$ can be made SPR. In [53], it was shown that for a combustion system, for 20% parametric uncertainties, the pressure oscillations decay. For the same level of uncertainties, controllers with fixed parameters were shown to result in destabilization.

3.3.2 Adaptive Time-delay Control

The structure of the model, and in particular, the fact that the relative degree was two was utilized in the above to design a stable adaptive controller. The same fact was also used in designing an adaptive controller in the presence of the delay in [14, 54]. In [14], the following adaptive controller was shown to be stabilizing:

$$\begin{aligned}\dot{\omega}_1 &= \Lambda_0\omega_1 + \ell u(t-\tau)\\ \dot{\omega}_2 &= \Lambda_0\omega_2 + \ell y(t)\\ u &= \theta_1^T(t)\omega_1 + \theta_2^T(t)\omega_2 + r(t) + \overline{\lambda}^T(t)\overline{u}(t)\\ \dot{\theta}(t) &= -y(t)\omega(t-\tau)\end{aligned} \tag{40}$$

where Λ_0 is an $n \times n$ matrix, (Λ_0, ℓ) is controllable, $\theta = [\theta_1^T, \theta_2^T, \overline{\lambda}^T]^T$, $\omega = [\omega_1^T, \omega_2^T, \overline{u}^T]^T$, and $\overline{u}_i$, the ith element of the vector $\overline{u}(t)$ is the ith sample of $u(t)$ in the interval $[t-\tau, t)$, $i = 1, \ldots, p$. The above controller can be guaranteed to stabilize the plant in (37) for any τ, for all initial conditions in a compact domain, whose size increases as τ approaches zero. Further extensions to the case when a saturation constraint is present in the control input have also been carried out [55].

3.3.3 Simulation and Experimental Results

An uncontrolled combustor simulated at an operating condition of Mach number = 0.08, equivalence ratio = 0.7 and mean temperature of 287.6 K, which corresponds to a 250 kW combustor considered in [56]. A low-order controller was implemented as in (40). A background noise that is 10% of the mean pressure was added to test the robustness properties of the controller and a time-delay of 7.5 ms was included in the plant. The advantage of the time-delay controller is clearly illustrated in the figure. It was also observed that a controller that altogether neglects the presence of the delay was ineffective at this delay value. The same controller was also implemented experimentally in a Rijke tube, where time-delay was introduced through software. The adaptive time-delay controller was shown to result in successful pressure suppression with a 50% variation in the unstable frequency, and a time-delay that is four times the period of instability [57].

3.3.4 Extremum-seeking Control

In [8], an adaptive version of the phase-shift controller based on an extremum-seeking approach has been proposed for the control of a single nozzle sector rig. The idea behind the extremum approach is the following: Suppose the pressure dynamics is given by

$$\dot{p} = -\alpha(\theta_c)(p - g(\theta_c))$$

and the idea is to determine θ_c^* where $g(\cdot)$ achieves a local minimum. By choosing $\theta_c(t) = \theta_m(t) + a\sin\omega t$, and determining algorithms for adjusting θ_m on-line, θ_m can be driven to a locally optimal value [58]. The same idea can be extended for the pressure dynamics in a general combustor.

In [8], the model in (27) is used as the starting point, which is simplified using the method of averaging. An Extended Kalman Filter (EKF) is used as an observer in order to track the frequency on-line so that variations in the frequency of the pressure oscillations can be reliably tracked. Using the resulting model and observer, the extremum seeking controller is used and it is shown that satisfactory pressure suppression can be achieved over a range of operating conditions in a full-scale engine.

3.3.5 Observer-based Control

Instead of an EKF, a different nonlinear observer is constructed in [59] where the sustained pressure oscillations are modeled as a sinusoidal signal whose parameters including amplitude, frequency, and phase are determined using the observer. The resulting parameters are then used to design a phase-shift controller. The following is a summary of the approach in [59]: The combustor pressure $p(t)$ is modeled as a sum of N sinusoids

$$p(t) = \sum_{i=1}^{N} (S_n \sin \Omega_n t + C_n \cos \Omega_n t)$$

For the case when $N = 1$, dropping the subscripts, the quantities Ω, S, and C are identified using the equations

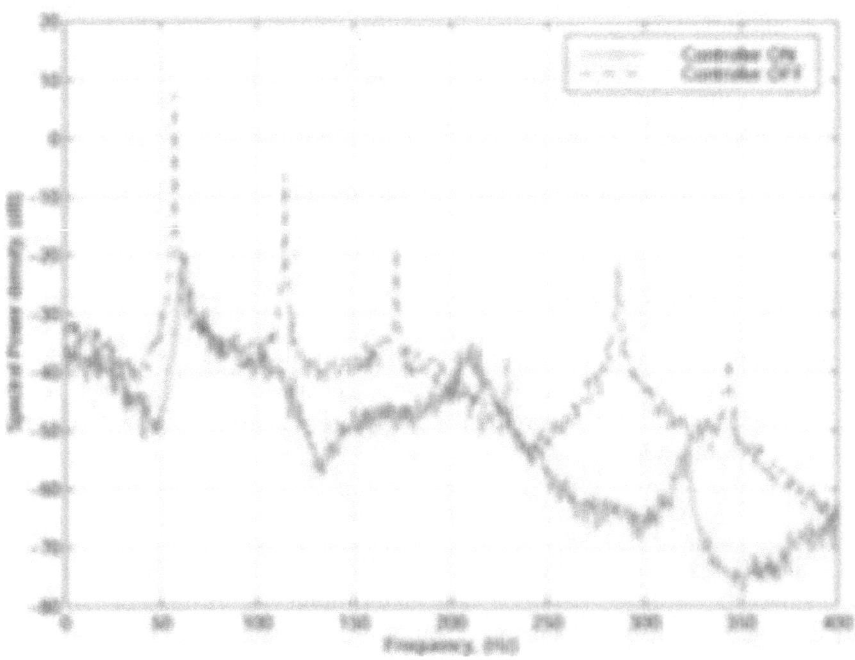

Fig. 5. Response of the controlled combustor with a time-delay of 7.5 ms in the input signal

$$\begin{aligned}
\dot{\widehat{\Omega}} &= \frac{\widehat{\Omega}^2}{2\pi\xi}\left[\sqrt{1+\frac{p(t)-p(t-\tau)}{\pi(\widehat{S}\cos(\widehat{\Omega}t)-\widehat{C}\sin(\widehat{\Omega}t))}}-1\right] \\
\dot{\widehat{S}} &= \frac{\widehat{\Omega}}{\pi}\left[F(t)-F(t-\tau)\right]\sin(\widehat{\Omega}(t-\tau_{i-1})) \\
\dot{\widehat{C}} &= \frac{\widehat{\Omega}}{\pi}\left[F(t)-F(t-\tau)\right]\cos(\widehat{\Omega}(t-\tau_{i-1}))
\end{aligned}$$

where $\tau = 2\pi/\widehat{\Omega}$, $\tau_i = 2\pi/\widehat{\Omega}(t_i)$, and the values of $\widehat{C}$ and $\widehat{S}$ are reset at t_i, $i = 1, \ldots$, where $t_i = t_{i-1} + 2\tau_i$, using functions $A(t)$ and $B(t)$ which are adjusted as

$$\begin{aligned}
\dot{\widehat{A}} &= \frac{\widehat{\Omega}}{\pi}F(t)\sin(\widehat{\Omega}(t-\tau_{i-1})) \\
\dot{\widehat{B}} &= \frac{\widehat{\Omega}}{\pi}F(t)\cos(\widehat{\Omega}(t-\tau_{i-1}))
\end{aligned}$$

While the analytical properties of this algorithm have not been studied at length, simulation studies show that for a large number of initial conditions, these estimates indeed converge to their true values. This algorithm has also been experimentally validated successfully in [7]. Using a fuel injector with a bandwidth of about 1 kHz [60], a phase-shift control algorithm in conjunction with the observer described above was implemented in a semi-scale gas turbine and shown to result in a four-fold reduction in the pressure oscillation level [7].

3.4 Neural Control

An alternate approach for controlling the combustion given the uncertain nature of the nonlinearities is to use a nonlinear controller that adapts to the uncertainties. One such nonlinear controller is based on neural networks where the latter seeks to identify the requisite nonlinear controller through training. The idea behind the neural controller is briefly described below.

Suppose the plant to be controlled is given by a vector differential equation

$$\dot{x} = f(x) + Bu \qquad y = Cx$$

and the stabilizing controller is given by

$$u = \gamma(x)$$

In general, the stabilizing controller $\gamma(\,\cdot\,)$ cannot be determined either because f is unknown or if f has a complex structure that does not satisfy the necessary involutive conditions [61]. In these cases, a neural controller may stabilize the system if it is appropriately constructed. One such method can be found in [62] which is described below: Construct the neural controller as

$$u = N(x, W) \tag{41}$$

where W are the weights of the neural network to be adjusted, and N is any neural network that has a universal approximator property [62], including multilayered perceptrons and radial basis functions. The weights W are adjusted as

$$\Delta W_j = -\rho \frac{\partial J}{\partial W_j} \qquad \rho > 0$$

such that the cost function J defined below is minimized:

$$\begin{aligned} J &= \sum_{i=1}^{M} \Delta V_{e_i}^2 \qquad \Delta V_{e_i} = \dot{V}_i - \dot{V}_{d_i} \\ \dot{V} &= 2x^T P[f(x) + BN(x, W)] \qquad \dot{V}_d = -x^T Q(x) x \\ Q(x) &= 1 + e^{-a\|x\|^2} \end{aligned} \tag{42}$$

P is the solution of the Lyapunov equation of the linearized plant, M is the number of nodes, and V is the Lyapunov function for the linearized plant.

The choice of J as in (42) is motivated by the following: The combustion system is strongly unstable at the origin with the degree of instability decreasing as the system approaches the limit-cycle. This is reflected in the choice of $Q(x)$ which becomes large as x approaches zero.

Simulation Results: Using the nonlinear model as in (22), and acoustics as in (2), a neural controller as in (41) was simulated, with $a = 0.17$, in parallel to an LQG-LTR controller. The resulting pressure responses are shown in Fig. 6 which shows a 50% improvement in the settling time with the neural controller compared to the linear controller [26].

4 Summary

In this paper, highlights of the area of modeling of combustion dynamics and model-based control strategies were presented. Linear models that capture the instability as well as nonlinear models that characterize limit-cycle behavior were described. In the latter case, it was shown that models stemming from distinct physical mechanisms can all lead to sustained pressure oscillations. Various model-based control strategies were also described in the paper. These include linear control methods where the underlying nonlinearities were ignored, time-delay control methods, adaptive control strategies that accommodated time-delays and parametric uncertainties, and neural network based strategies that suitably compensated for the presence of limit-cycle nonlinearities.

Active combustion control is a very viable and feasible technology, as illustrated by a range of experiments from a 1 kW benchtop rig to a 265 MW

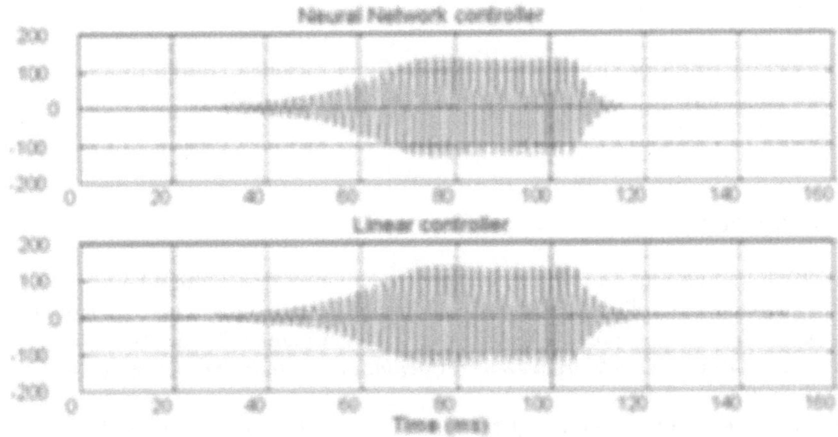

Fig. 6. Response of the combustor using a neural network-based controller; the correspnding response using a linear LQG-LTR controller is shown for comparison

industrial rig. Results accruing from model-based control strategies illustrate that an order of magnitude improvement in the pressure response can be obtained over their empirical counterparts. Buoyed by these results, the time is quite appropriate to investigate the realization of uniform performance over a range of equivalence ratios and flow-rates in the presence of variations in fuel, combustion, boundary conditions, and system parameters.

While better models are being generated for heat-release dynamics, their interaction with the underlying hydrodynamics remains to be quantified. Vortex-driven heat-release dynamics is an area where much work remains to be carried out. Systematic methods for deriving control-oriented models from detailed computational models of the reactive flow field in a combustor are needed. Further investigations are also required in synthesizing actuators with the requisite bandwidth and can sustain hazardous conditions of high pressure and temperature typically encountered in practice.

Combustion instability suppression is but one objective of combustion control. Maintaining high premixedness, low NOx levels, complete combustion, and control of pattern factor are some of the other typical concomitant requirements in applications such as gas turbines, afterburners, and ramjet engines. This requires the judicious use of multiple and distributed actuators that function synergistically with the combustion dynamics and provides simultaneous optimization of various objectives in the next generation engines and power systems.

Acknowledgements

This work is sponsored in part by the National Science Foundation, contract no. ECS 9713415, and in part by the Office of Naval Research, contract no. N00014-99-1-0448.

References

[1] J.W.S. Rayleigh. *The Theory of Sound*, volume 2. Dover, New York, 1945.

[2] B.T. Chu. "Stability of systems containing a heat source–The Rayleigh criterion". Technical report, NASA Research Memorandum RN 56D27, 1956.

[3] A.A. Putnam. *Combustion Driven Oscillations in Industry.* American Elsevier Pub. Co., NY, 1971.

[4] P.J. Dines. *Active control of flame noise.* PhD thesis, Cambridge University, England, 1984.

[5] M.A. Heckl. Active control of the noise from a Rijke tube. In G. Comte-Bellot and J.E. Flowers Williams, editors, *Aero- and Hydro-Acoustics*, pages 211–216. Springer, Berlin Heidelberg, Berlin Heidelberg, 1986.

[6] J. Hermann, A. Orthmann, S. Hoffmann, and P. Berenbrink. Combination of active instability control and passive measures to prevent combustion instabilities in a 260 mw heavy duty gas turbine. In *NATO RTO/AVT Symposium on Active Control Technology for Enhanced Performance in Land, Air, and Sea Vehicles*, Braunschweig, Germany, May 2000.

[7] S.S. Sattinger, Y. Neumeier, A. Nabi, B.T. Zinn, D.J. Amos, and D.D. Darling. Sub-scale demonstration of the active feedback control of gas-turbine combustion instabilities. *ASME Journal of Engineering for Gas Turbines and Power*, 122:262–268, 2000.

[8] A. Banaszuk, Y. Zhang, and C. Jacobson. Active control of combustion instabilities in gas turbine engines for low emissions. Part II: Adaptive control algorithm development, demonstration, and performance limitations. In *NATO RTO/AVT Symposium on Active Control Technology for Enhanced Performance in Land, Air, and Sea Vehicles*, Braunschweig, Germany, May 2000.

[9] A.M. Annaswamy and A.F. Ghoniem. Active combustion control: Theory and practice. *IEEE Control Systems Magazine*, 2001 (under review).

[10] F.E.C. Culick. Nonlinear behavior of acoustic waves in combustion chambers. *Acta Astronautica*, 3:715–756, 1976.

[11] F.E.C. Culick. "Combustion instabilities in liquid-fueled propulsion systems - An Overview". In *AGARD Conference Proceedings, paper 1, 450, The* 72nd *(B) Propulsion and Energetics Panel Specialists Meeting*, 1988.

[12] A.M. Annaswamy, M. Fleifil, J.P. Hathout, and A.F. Ghoniem. Impact of linear coupling on the design of active controllers for thermoacoustic instability. *Combust. Sci. Tech.*, 128:131–180, 1997.

[13] A.P. Dowling. "A kinematic model of of a ducted flame". *Journal of Fluid Mechanics*, 394:51–72, 1999.

[14] S. Evesque, A.P. Dowling, and A.M. Annaswamy. Adaptive algorithms for control of combustion. In *NATO RTO/AVT Symposium on Active Control Technology for Enhanced Performance in Land, Air, and Sea Vehicles*, Braunschweig, Germany, May 2000.

[15] J.P. Hathout, A.M. Annaswamy, and A.F. Ghoniem. Modeling and control of combustion instability using fuel injection. In *NATO RTO/AVT Symposium on Active Control Technology for Enhanced Performance in Land, Air, and Sea Vehicles*, Braunschweig, Germany, May 2000.

[16] M. Fleifil, A.M. Annaswamy, Z. Ghoniem, and A.F. Ghoniem. Response of a laminar premixed flame to flow oscillations: A kinematic model and thermoacoustic instability result. *Combust. Flame*, 106:487–510, 1996.

[17] Y.B. Zeldovich, G.I. Barenblatt, V.B. Librovich, and G.M. Makhviladze. *The mathematical theory of combustion and explosions.* Consultant Bureaum, New York, NY, 1985.

[18] M.C. Janus and G.A. Richards. Results of a model for premixed combustion oscillations. In *Proceedings of the 1996 AFRC International Symposium*, Baltimore, MD, 1996.

[19] T. Lieuwen and B.T. Zinn. "The role of equivalence ratio oscillations in driving combustion instabilities in low $N0_x$ gas turbines". *The Twenty Seventh International Symposium on Combustion*, pages 1809–1816, 1998.

[20] S. Park, A.M. Annaswamy, and A.F. Ghoniem. Heat release dynamics modeling of kinetically controlled burning. In *38th AIAA Aerospace Sciences Meeting and Exhibit*, Reno, NV, 2001.

[21] A.A. Peracchio and W. Proscia. Nonlinear heat release/acoustic model for thermoacoustic instability in lean premixed combustors. In *ASME Gas Turbine and Aerospace Congress*, Sweden, 1998.

[22] J.P. Hathout, M. Fleifil, A.M. Annaswamy, and A.F. Ghoniem. Heat-release actuation for control of mixture-inhomogeneity-driven combustion instability. In *28th International Symposium on Combustion*, University of Edinburgh, Scotland, July 2000.

[23] M. Fleifil, J.P. Hathout, A.M. Annaswamy, and A.F. Ghoniem. The origin of secondary peaks with active control of thermoacoustic instability. *Combustion, Science, and Technology*, 133:227–265, 1998.

[24] A.M. Annaswamy, M. Fleifil, J. Rumsey, J.P. Hathout, R. Prasanth, and A.F. Ghoniem. Thermoacoustic instability: Model-based optimal control designs and experimental validation. *IEEE Transactions on Control Systems Technology*, 8(6):905–918, November 2000.

[25] S. Murugappan, S. Park, A.M. Annaswamy, A.F. Ghoniem, S. Acharya, and T. Allgood. Optimal control of a swirl stabilized spray combustor using system identification approach. In *38th AIAA Aerospace Sciences Meeting and Exhibit*, Reno, NV, 2001.

[26] M. Fleifil, A.M.J.P. Annaswamy, and A.F. Ghoniem. "A physically based nonlinear model of combustion instability and active control". In *Proceedings of the Conference on Control Applications*, Trieste, Italy, August 1998.

[27] J. Rumsey, M. Fleifil, A.M. Annaswamy, J.P. Hathout, and A.F. Ghoniem. "Low-order nonlinear models of thermoacoustic instabilities and linear model-based control". In *Proceedings of the Conference on Control Applications*, Trieste, Italy, August 1998.

[28] T. Lieuwen and B.T. Zinn. "Experimental investigation of limit cycle oscillations in an unstable gas turbine combustor". In *AIAA 2000-0707, 38th AIAA Aerospace Sciences Meeting*, Reno, NV, January 2000.

[29] V. Yang and F.E.C. Culick. "Nonlinear analysis of pressure oscillations in ramjet engines". *AIAA-86-0001*, 1986.

[30] F.E.C. Culick. "Some recent results for nonlinear acoustics in combustion chambers". *AIAA*, 32:146–169, 1994.
[31] G.A. Richards, M.C. Yip, and E.H. Rawlins. Control of flame oscillations with equivalence ratio modulation. *Journal of Propulsion and Power*, 15:232–240, 1999.
[32] G. Isella, C. Seywert, F.E.C. Culick, and E.E. Zukoski. A further note on active control of combustion instabilities based on hysteresis. *Short Communication, Combustion, Science, and Technology*, 126:381–388, 1997.
[33] R. Prasanth, A.M. Annaswamy, J.P. Hathout, and A.F. Ghoniem. "When do open-loop strategies for combustion control work?". *36th AIAA/ASME/SAE/ASEE Joint Propulsion Conference, paper no. 2000-3350*, 2000.
[34] A.P. Dowling. Nonlinear acoustically-coupled combustion oscillations. *2nd AIAA/CEAS Aeroacoustics Conference*, May 6-8 1996.
[35] P. Shih-I. *Fluid Dynamics of Jets*. D. Van Nostrand Co., Inc., USA, 1954.
[36] S.B. Park, J.P. Hathout, M. Fleifil, A.M. Annaswamy, and A.F. Ghoniem. Nonlinear flame dynamics and limit cycles in a combustor. Technical report, Adaptive Control Laboratory, MIT, Cambridge, MA, April 2001.
[37] G.J. Bloxsidge, A.P. Dowling, N. Hooper, and P.J. Langhorne. "Active control of reheat buzz.". *AIAA Journal*, 26, No. 7, July 1989.
[38] J.E. Tierno and J.C. Doyle. Multimode active stabilization of a Rijke tube. In *DSC-Vol. 38*. ASME Winter Annual Meeting, 1992.
[39] R. Prasanth, R.K. Mehra, and A.M. Annaswamy. A system identification model of the mit laminar combustor and model based control. Technical report, Adaptive Control Laboratory, Department of Mechanical Engineering, MIT, Cambridge, MA, 1999.
[40] S. Koshigoe, T. Komatsuzaki, and V. Yang. Adaptive control of combustion instability with on-line system identification. *Journal of Propulsion and Power*, 15:383–389, 1999.
[41] B.J. Brunell. "A system identification approach to active control of thermoacoustic instabilities". Master's thesis, Department of Mechanical Engineering, M.I.T., Cambridge, MA., 1996.
[42] K. Yu, K.J. Wilson, and K.C. Schadow. Scale-up experiments on liquid-fueled active combustion control. In *34th AIAA/ASME/SAE/ASEE Joint Propulsion Conference*, pages AIAA 98–3211, Cleveland, OH, 1998.
[43] J. Rumsey, M. Fleifil, A.M. Annaswamy, and A.F. Ghoniem. "Low order nonlinear models for thermoacoustic instability". In *Proceedings of the Conference on Control Applications*, Trieste, Italy, August 1998.
[44] G. Stein and M. Athans. The LQG/LTR procedure for multivariable feedback control design. *IEEE Transactions on Automatic Control*, 32:105–114, 1987.
[45] S. Murugappan, S. Acharya, E. Gutmark, and T. Messine. "Active control of combustion instabilities in spray combustion with swirl". *AIAA-2000-1026, 38th AIAA Aerospace Sciences Meeting, Reno, NV*, 2000.
[46] O.J. Smith. "A controller to overcome dead time". *ISA Journal*, 6, 1959.
[47] A.Z. Manitius and A.W. Olbrot. Finite spectrum assignement problem for systems with delays. *IEEE Transactions on Automatic Control*, AC-24 no. 4, 1979.
[48] K. Ichikawa. Frequency-domain pole assignement and exact model-matching for delay systems. *Int. J. Control*, 41:1015–1024, 1985.

[49] S.I. Niculescu and A.M. Annaswamy. A simple adaptive controller for positive-real systems with time-delay. In *Proc. American Control Conference*, Chicago, IL, February 2000.
[50] B. Widrow and S.D. Stearns. *Adaptive signal processing.* Prentice-Hall, Inc., Englewood Cliffs, N.J., 1985.
[51] G. Billoud, M.A. Galland, C. Huynh Huu, and S. Candel. Adaptive active control of combustion instabilities. *Combust. Sci. and Tech.*, 81:257–283, 1992.
[52] A. Kemal and C.T. Bowman. Active adaptive control of combustion. In *Proceedings of the IEEE Conference on Control Applications*, pages 667–672, Albany, NY, 1995.
[53] A.M. Annaswamy, O. El-Rifai, M. Fleifil, J.P. Hathout, and A.F. Ghoniem. A model-based self-tuning controller for thermoacoustic instability. *Combustion Science and Technology*, 135:213–240, 1998.
[54] S. Niculescu and A.M. Annaswamy. A simple adaptive controller for positive-real systems with time-delay. In *The American Controller Conference*, Chicago, IL, 2000.
[55] S. Evesque, A.P. Dowling, and A.M. Annaswamy. Adaptive combustion instability control with saturation. In *37th AIAA/ASME/SAE/ASEE Joint Propulsion Conference (to appear)*, Salt Lake City, Utah, July 2001.
[56] P.J. Langhorne, A.P. Dowling, and N. Hooper. Practical active control system for combustion oscillations. *Journal of Propulsion and Power*, 6(3):324–333, 1990.
[57] S. Evesque. *Adaptive Control of Combustion Oscillations.* PhD thesis, University of Cambridge, Cambridge, UK, November 2000.
[58] M. Krstić. Performance improvement and limitations in extremum seeking control. *Systems and Control Letters*, 39:313–326, 2000.
[59] Y. Neumeier, N. Markopoulos, and B.T. Zinn. A procedure for real-time mode decomposition, observation, and prediction for active control of combustion instabilities. In *Proceedings of the IEEE Conference on Control Applications*, Hartford, CT, 1997.
[60] Y. Neumeier. Investigation of the open loop performance of an active-control system using a fuel injector actuator. In *32nd AIAA/ASME Joint Propulsion Conference*, Lake Buena Vista, FL, 1996.
[61] A. Isidori. *Nonlinear Control Systems: An Introduction.* Springer-Verlag, New York, NY, 1985.
[62] S. Yu and A.M. Annaswamy. Stable neural controllers for nonlinear dynamic systems. *Automatica*, 34:669–679, May 1998.

Control of Acoustics

Stephen Elliott

Institute of Sound and Vibration Research, University of Southampton, Highfield, Southampton SO17 1BJ, UK

Abstract. The active control of sound transmission in a duct is used to illustrate both the physical principles of control and its practical implementation. The control of more complicated acoustic fields, such as those inside enclosures, is then described, with particular reference to the reduction of noise inside aircraft. Finally, the potential for controlling sound generation at source is discussed, using flow-excited cavity oscillations as an example.

1 Introduction

Control can potentially be applied either to the generation of an acoustic disturbance, or to its transmission as sound. The control of sound transmission is considered in the first part of this chapter, starting with the control of plane acoustic waves in a one-dimensional duct to illustrate the physical principles of reflection and absorption. The formulation of this problem in terms familiar from automatic control is considered next, in which the availability of a time-advanced reference signal leads to the use of a feedforward controller. In practice the system under control is not exactly known and the feedforward controller is generally made adaptive, in which case the control system becomes closed-loop with provable stability and performance properties.

Although commercial systems using these principles are available for the control of low frequency sound in air conditioning systems, for example, recent research has focused on the active control of multiple modes in short ducts because of the interest in reducing the inlet noise from turbofan engines while an aircraft is landing.

The active control of internal noise in propeller aircraft is also an important application area. The physical principles behind the control of such an enclosed soundfield is described in the following section, in which it is emphasised how the modal characteristics of such an enclosure impose a fundamental high-frequency limit on the use of active control.

Both the control of sound in ducts and in enclosures are generally linear problems for the applications described above. Sound is often generated

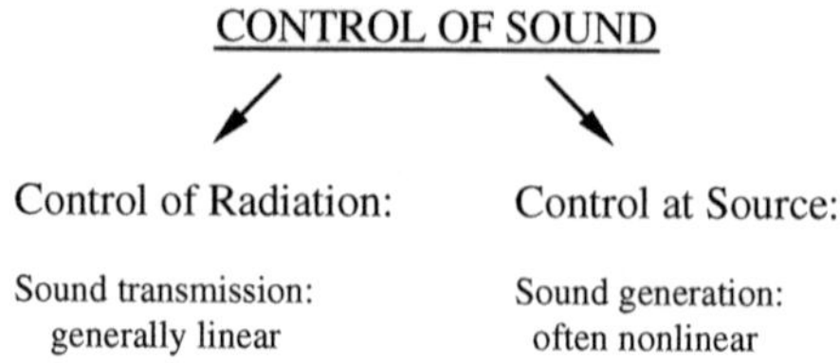

Fig. 1. Structure of the chapter

by a nonlinear process, however, and the active control of sound at source must take this nonlinearity into account. The technology for such a control approach is far less developed, but because of the extreme sensitivity of many nonlinear processes to small perturbations, it does hold the promise in the long term of being able to suppress powerful sources of sound with modest control actuators. Potential applications in controlling the flow-induced sound in cavities and in the control of suppressor surge are very briefly discussed. The distinction between the control of sound transmission, or radiation, and sound generation, i.e. control at source, is indicated in Fig. 1.

2 Control of Plane Waves in Ducts

At frequencies below the cut-on frequency, only plane sound waves can propagate in a duct, and the sound transmission problem becomes essentially one-dimensional. Provided the acoustic pressure $p(x,t)$ is always much less than the ambient pressure, then away from any sources it obeys the one-dimensional homogeneous wave equation.

$$\frac{\partial^2 p(x,t)}{\partial x^2} = \frac{1}{c_o^2}\frac{\partial^2 p(x,t)}{\partial t^2}, \tag{1}$$

where c_o is the speed of sound. The waveform of a sinusoidal acoustic wave of frequency ω can be represented as the real part of the complex pressure

$$p(x) = p_+ \, e^{-jkx} + p_- \, e^{+jkx}, \tag{2}$$

where $k = w/c_o$ is the wavenumber, and p_+ and p_- are the complex amplitudes of the waves travelling in the positive and negative x directions, i.e. downstream and upstream.

If an incident downstream wave of amplitude A is present in the duct, together with a secondary monopole acoustic source at $x = 0$, which generates upstream and downstream waves of amplitude B, then the total complex pressure is

$$p(x) = A\, e^{-jkx} + B\, e^{-jkx} \qquad \text{for } x > 0 \tag{3}$$

and

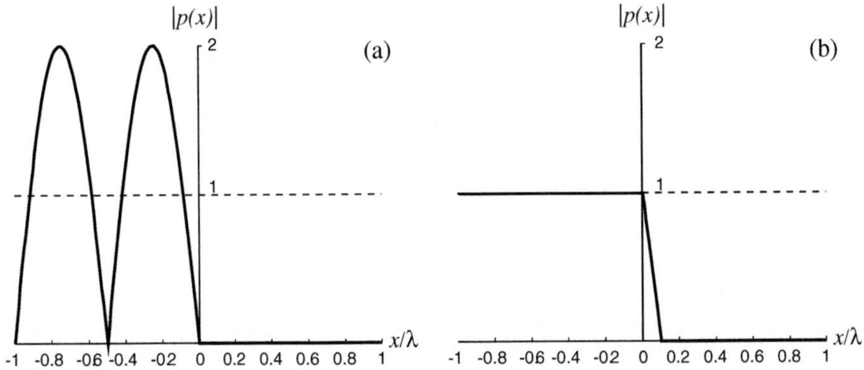

Fig. 2. The distribution of pressure amplitude in a duct carrying an incident plane wave in the positive x direction alone (dotted line) and when the sound transmission is actively controlled by a single secondary source, which reflects the incident wave, (**a**), and by a pair of secondary sources, which absorb the incident wave, (**b**)

$$p(x) = A\, e^{-jkx} + B\, e^{+jkx} \qquad \text{for } x < 0\,. \tag{4}$$

The amplitude and phase of the secondary source can be adjusted so that the total pressure everywhere downstream in the duct is zero, as originally suggested by Lueg (1936) [1], in which case $B = -A$ and so

$$p(x) = 0 \qquad \text{for } x < 0\,, \tag{5}$$

in which case

$$p(x) = -2_j A \sin kx \qquad x < 0\,. \tag{6}$$

The single secondary source has effectively created a pressure-release condition at $x = 0$, which reflects the incident acoustic wave back down the duct and generates perfect cancellation on one side and a standing wave on the other. Figure 2(a) illustrates the distribution of the pressure amplitude in this case.

If two controllable monopole sources are used as secondary sources in the duct then not only can the incident wave be cancelled in the downstream direction, but the reflected wave in the upstream direction can be suppressed (Swinbank, 1973) [2], in which case the secondary source array absorbs the energy of the incident soundwave, and the distribution of the pressure amplitude is as shown in Fig. 2(b).

3 Controller Design and Implementation

In practice the waveform of the incident soundwave must be measured by a microphone acting as a reference sensor. The residual pressure may be measured by another microphone acting as an error sensor, as shown in Fig. 3(a).

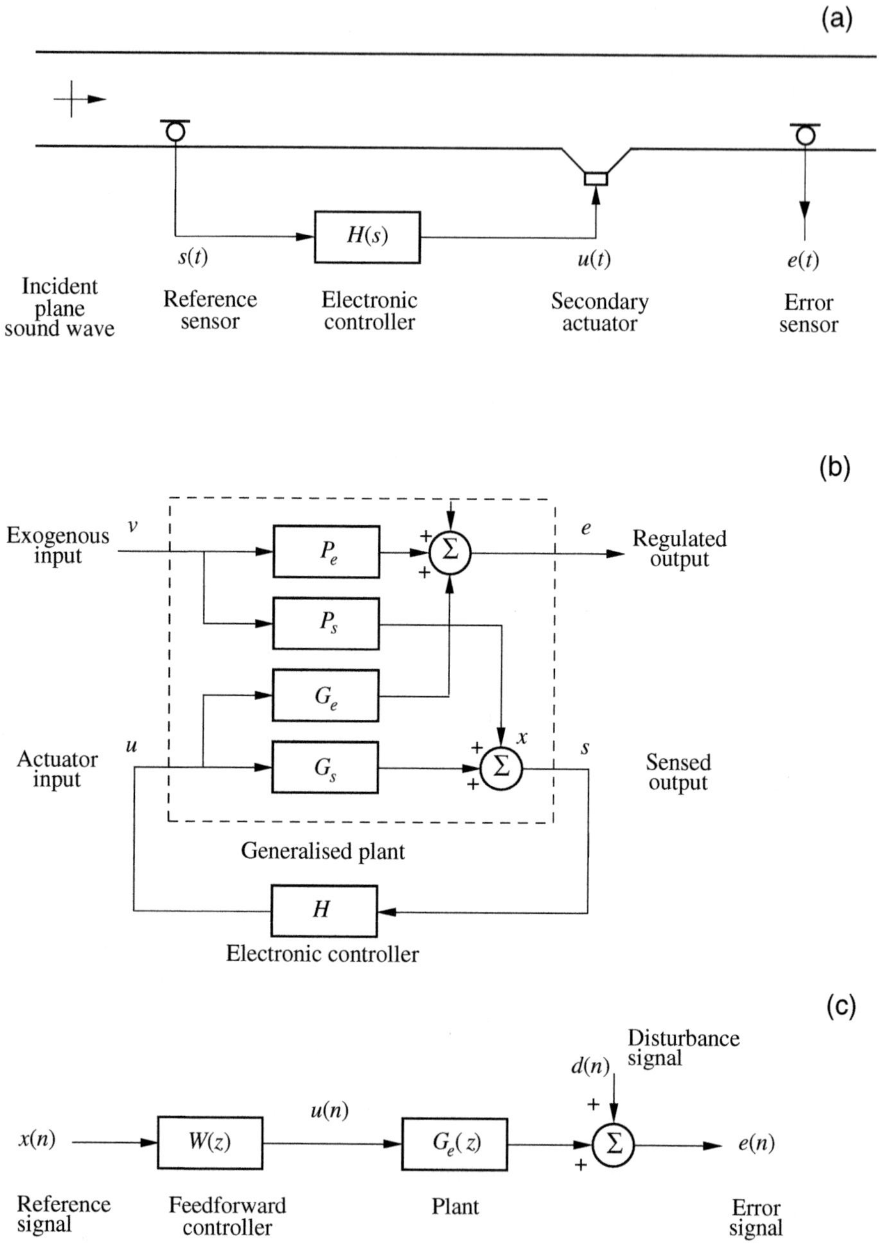

Fig. 3. A feedforward system for control of sound in a duct (**a**), its equivalent block diagram (**b**), and its simplified block diagram assuming the controller, H, internally cancels the path G_s

The system under control, as shown in Fig. 3(b), now has two inputs, one generating the incident wave, which is exogeneous, and the other driving the secondary actuator, which is the control input, and two outputs, one from the reference sensor, the sensed output, and one from the error sensor, the regulated output (Clark, et al., 1998) [3]. The individual transfer functions which can be measured are denoted G_e and G_s and those from the unobservable exogenous input are denoted P_e and P_s.

Assuming that the electronic controller contains an internal model of the electroacoustic feedback path from the secondary actuator to the reference sensor, which is used to cancel the effects of G_s in Fig. 3(b), the block diagram can be written in the simplified feedforward form shown in Fig. 3(c).

Knowing the plant response, $G_e(z)$, and the spectral properties of $e(n)$ and $d(n)$, the transfer function of the feedforward controller which minimises the mean-square value of the error can be calculated using Weiner filtering theory, even when measurement noise is present in the output of the reference sensor, which degrades the performance (Roure, 1985) [4]. In practice the plant response and the spectral properties of $e(n)$ and $d(n)$ will change with time and an adaptive algorithm is generally used to maintain good control performance. This adaptation ensures that the control system is "closed loop", and the algorithms can be analysed to give clear conditions for robustness and performance [6].

4 Control of Higher-order Modes in Ducts

At higher excitation frequencies, when the acoustic wavelength is no longer small compared with the dimensions of the duct cross section, higher order modes can propagate in the duct. The number of higher order modes which can propagate increases with frequency, as illustrated for a circular duct in Fig. 4, and eventually is proportional to the square of the excitation. In general it would require N secondary sources to control N modes in the duct, provided the secondary sources could independently excite each of these modes.

Several authors have considered the active control of higher order modes in ducts, e.g. Fedorynk (1975) [7], Zander and Hansen (1992) [8]. There is currently an increasing interest in the control of tonal noise propagating as higher order modes because of the potential application in controlling the fan tones radiated from the inlet of aircraft engines, particularly as the aircraft is coming in to land (see, for example, Burdisso et al., 1993) [9]. Unfortunately, at the frequencies of these tones there may be 120 higher-order modes which can propagate in the short inlet duct of an aircraft engine (Nelson, 2000) [11], and a control system to control all of these modes would be very complicated. Joseph et al. (1996) [10] have shown, however, that only a much smaller number of modes need be actively controlled if only the sound radiated in particular directions, such as on the ground, needs to be reduced. Nelson (2000) [11] also suggests a way in which it may be possible to control these fan tones

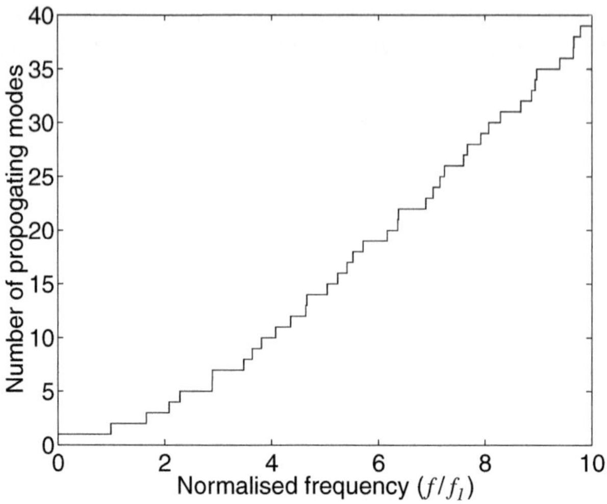

Fig. 4. The number of higher-order modes which can propagate in a circular duct as a function of excitation frequency, f, normalised to the cut-on frequency of the first higher-order mode, f_1

by actively distorting the inlet flow conditions and thus interfering with the sound generation mechanism.

5 Control of Sound in Enclosures

In most practical applications sound propagation is three-dimensional and the simple control mechanisms of reflection and absorption seen in the one-dimensional duct are complicated by geometric considerations. In this section we consider the further complication of having the soundfield enclosed by a hard boundary. An important application area for active control is the reduction of noise in vehicles, particularly aircraft. In order to keep the formulation simple we will assume that the soundfield is tonal and continue to use the complex pressure at a single frequency, ω, which may now be described in modal form as

$$p(\boldsymbol{x},\omega) = \sum_{n=0}^{\infty} a_n(\omega)\Psi_n(\boldsymbol{x}) \; , \tag{7}$$

where $\boldsymbol{x}$ is the position vector, $a_n(\omega)$ is the amplitude of the n-th acoustic mode and $\Psi_n(\boldsymbol{x})$ is its mode shape.

Although in principle an infinite number of modes must be used to describe the sound in the enclosure, because it is a distributed parameter system, the soundfield can always be approximated to arbitrary accuracy with a finite modal series. In the low frequency range, where active control is most effective, the modal description is a very efficient representation of the soundfield. Conventional passive noise control techniques also do not work very well in this low frequency region, unless very massive barriers or bulky absorbers are used, and so active control conveniently complements the effect of passive noise control techniques and can provide significant weight and space savings at low frequencies.

Two active control problems will be briefly considered: global control and local control. The objective of a global control system is to reduce the sound throughout the enclosure by adjusting the amplitudes and phases of a number of secondary sources, which are typically loudspeakers. The fundamental limits of such a strategy can be assessed by calculating the reductions which are possible in the total acoustic potential energy in the enclosure, which may be written as

$$E_p(\omega) = \frac{1}{4\rho_o c_o^2} \int_v |p(\mathbf{x},\omega)|^2 \, dV \, . \tag{8}$$

Assuming that the mode shapes are orthonormal, $E_p(\omega)$ is equal to the sum of the modulus squared mode amplitudes (Nelson and Elliott, 1992) [12]. Since the secondary sources linearly couple into each mode amplitudes, $E_p(\omega)$ is a quadratic function of the complex secondary source strengths, which has a unique global minimum. This minimum value of $E_p(\omega)$ provides a measure of the best performance that can be obtained in a global control system for a given distribution of secondary sources and a given excitation frequency. Figure 5(b), for example, shows the result of such a calculation for the levels of $E_p(\omega)$ in a computer model of an enclosure of dimensions $1.9 \times 1.1 \times 1.0$ m as shown in Fig. 5(a), which are approximately the conditions inside a small car (Elliott, 2001) [5]. The solid line in Fig. 5(b) shows the energy due to a primary monopole source in one corner of the enclosure, the dashed line the energy after it has been minimised using a single secondary monopole source in the opposite corner adjusted and the dot-dashed line after minimisation using 7 secondary sources placed at all the corners of the enclosure away from the primary. The first longitudinal resonance, at about 80 Hz, is significantly attenuated by the action of a single secondary source, but almost no reduction is achieved in the energy at about 160 Hz close to which three acoustic modes have their natural frequencies. This is to be expected, since in general a single source can only control a single mode,but even with 7 secondary sources a reduction in energy of only about 5dB is achieved at this excitation frequency and this reduction becomes less than 1dB at about 250 Hz.

The number of significantly excited modes can be estimated from the modal overlap, which is the average number of modes with natural frequencies within the half power bandwidth of a single mode, and the value of the modal overlap is plotted in Fig. 6 for an enclosure of the size used in the simulations for Fig. 5. The modal overlap rises above 7 at about 250 Hz, as expected, but at higher frequencies the modal overlap rises as the cube of the excitation

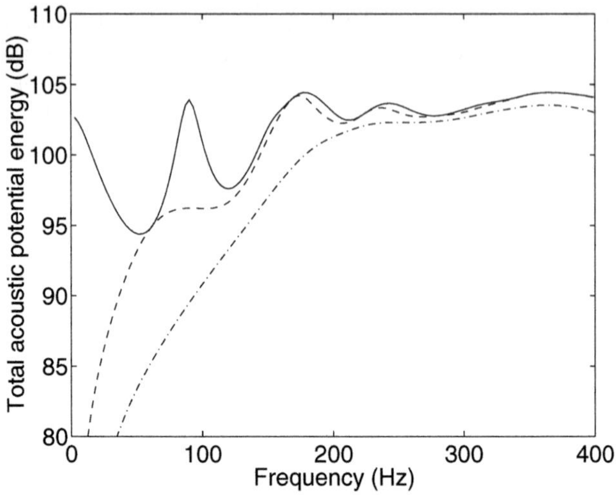

Fig. 5. The total acoustic potential energy in the enclosure as a function of excitation frequency, when driven by the primary source alone, solid line, when the energy is minimised using a single secondary source (dashed line) and when the energy is minimised using 7 secondary sources (dot-dashed line)

frequency. This feature provides a very clear upper frequency limit to global control with a reasonable number of secondary sources.

An alternative strategy to global control would be to only control the sound at specific locations in an enclosure, such as close to the ears of passengers

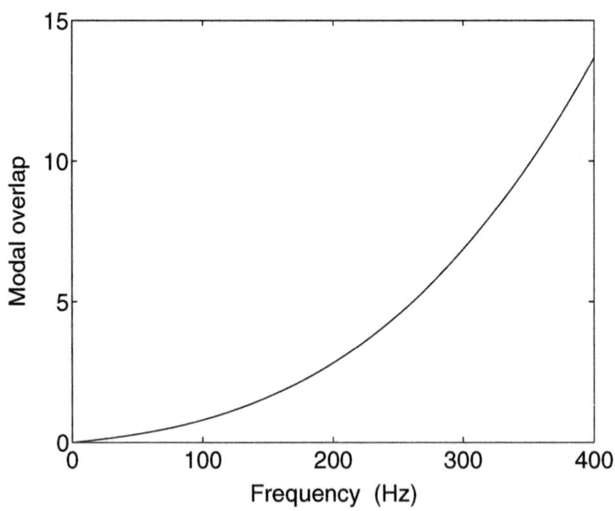

Fig. 6. The acoustic modal overlap for the enclosure shown in Fig. 5

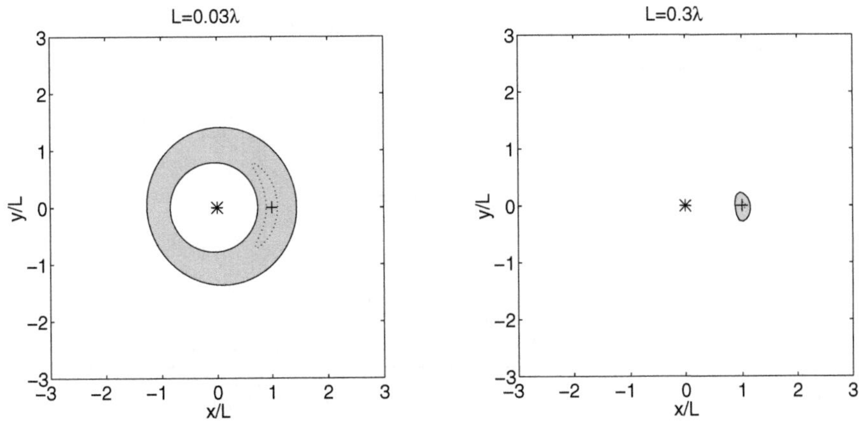

Fig. 7. The zone of quiet, within which the diffuse primary field is attenuated by more than 10dB for a local control system in which a monopole acoustic source at the origin is arranged to cancel the pressure at $x = L$ for two different excitation frequencies

in a vehicle. Such a local control strategy was originally suggested by Olson and May (1953) [13] who describe an active headrest using a feedback control system from a microphone to a closely-spaced loudspeaker acting as the secondary source. The acoustic performance of such a system depends on the detailed geometric arrangement of the headset and the position of the passenger's head, but some physical insight can be gained by considering simplified models. Figure 7, for example, shows a cross-section through the zone of quiet, within which the sound has been attenuated by at lest 10dB, generated when a diffuse primary soundfield is cancelled by an acoustic monopole at the origin (Garcia-Bonito and Elliott, 1996) [14] at the point $x = L$. The two graphs correspond to an excitation frequency for which L is much smaller than the acoustic wavelength, in which case a "shell" of quiet is generated around the secondary source, and to an excitation frequency for which L is of the order of the acoustic wavelength, in which case the zone of quiet is spherical with a diameter which is about one tenth of an acoustic wavelength (Elliott et al., 1988) [15].

6 Control of Sound at Source

In this section we briefly consider the control of sound generated aerodynamically. Many sources of sound are not aerodynamic but mechanical and the control of a structure's vibration to reduce sound radiation or active structural acoustic control (ASAC) can then be employed (Fuller et al., 1996) [15], but

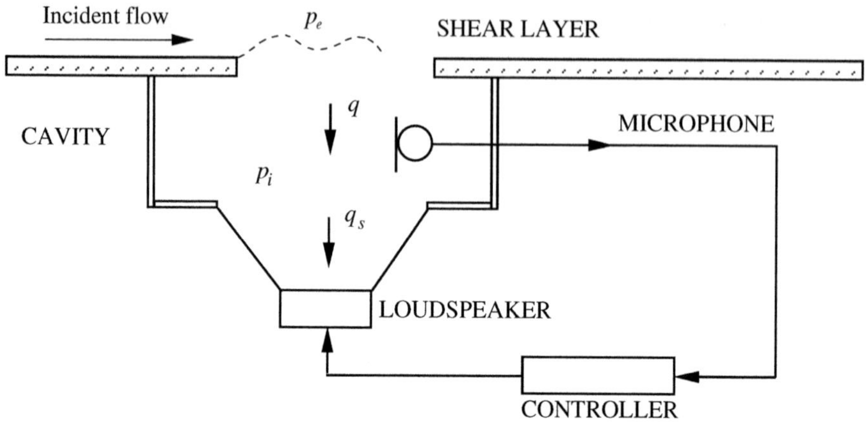

Fig. 8. Simplified block diagram of the feedback control of a flow-excited cavity, after Sunyach and Ffowcs Williams (1986)

the problem remains fundamentally linear. Aerodynamic sound sources are generally nonlinear, but in their simplest form they often start off as single-frequency oscillations, governed by a linear but unstable system of equations, and then grow into limit-cycle oscillations, which are governed by a nonlinear system of equations. An example of such a sound source is the flow-excited cavity illustrated in Fig. 8, in which a simple linear feedback controller has been assumed with a loudspeaker and microphone inside the cavity, as suggested by Sunyach and Ffowcs Williams (1986) [22] for example.

If the control system can stabilise the flow-induced pressure oscillations in the cavity while the amplitude of these oscillations are small, the system behaves almost linearly and a linear analysis can be used to illustrate the mechanism of control.

Assuming linear superposition and working in the frequency domain the governing equations for small amplitude oscillations in the cavity can be written in a form similar to that of Sunyach and Ffowcs Williams. The complex external pressure, p_e, can be written as

$$p_e = p_d - Z_r q \,, \tag{9}$$

where p_d is a disturbance, Z_r is the radiation impedance looking out of the cavity and q is the complex volume velocity of the flow into the cavity. The volume velocity drawn from the cavity by the loudspeaker and the feedback loop can be expressed as

$$q_s = Y_f p_i \,, \tag{10}$$

where Y_f is the response of the feedback controller and p_i is the pressure inside the cavity, which is assumed to be uniform and is given by

$$p_i = Z_c(q - q_s)\,, \tag{11}$$

where Z_c is the impedance of the cavity.

The pressure difference on either side of the cavity neck is finally assumed to be related to the volume velocity flowing through the neck by a complex frequency dependent impedance Z_n, so that

$$p_e - p_i = Z_n q. \tag{12}$$

Combining together these equations, the ratio of the internal pressure to the pressure disturbance can be written as

$$\frac{p_i}{p_d} = \frac{Z_c}{Z_c + (Z_n + Z_r)(1 + Z_c Y_f)}\,. \tag{13}$$

The combined effect of the acoustic neck impedance and the acoustic radiation impedance can be written as

$$Z_n + Z_r = R + j\omega L \tag{14}$$

where L is the total acoustic inertance and R is the total acoustic resistance. Sunyach and Ffowcs Williams discuss the fact that these quantities depend on the flow velocity over the neck and that the resistance R can become negative under certain flow conditions. Assuming also that $Z_c = 1/j\omega C$ where C is the acoustic compliance of the cavity, then

$$\frac{p_i}{p_d} = \frac{1}{1 + (R + j\omega L)(Y_f + j\omega C)}\,. \tag{15}$$

In the absence of feedback control, $Y_f = 0$, the cavity thus responds to a disturbance like a second order resonator and if the flow conditions are such that R is negative, the pressure oscillations at the natural frequency, $\omega_o = \sqrt{1/LC}$, will build up until limited by nonlinear effects not accounted for in the above analysis.

There are, however, a number of strategies which could be used in the feedback system to control this instability and thus not allow the pressure oscillations to build up to begin with. If $Y_f = -j\omega C$, for example, the loudspeaker motion would counteract the cavities compliance, in which case $p_i = p_d$ so that the system is stabilised. Alternatively, if Y_a had a positive real component which is greater than $-RC/L$ then the feedback loop would add sufficient damping to the cavity to suppress the oscillation. It is clear that the feedback controller can stabilise the system in a variety of ways and it is interesting to note that when an adaptive feedback controller was investigated for this problem by Billout et al. (1991) [17] they found that the error surface had multiple minima.

A practical application of such an adaptive feedback system for the control of sun roof flow oscillations in a car has been described by Stothers et al. (1993)

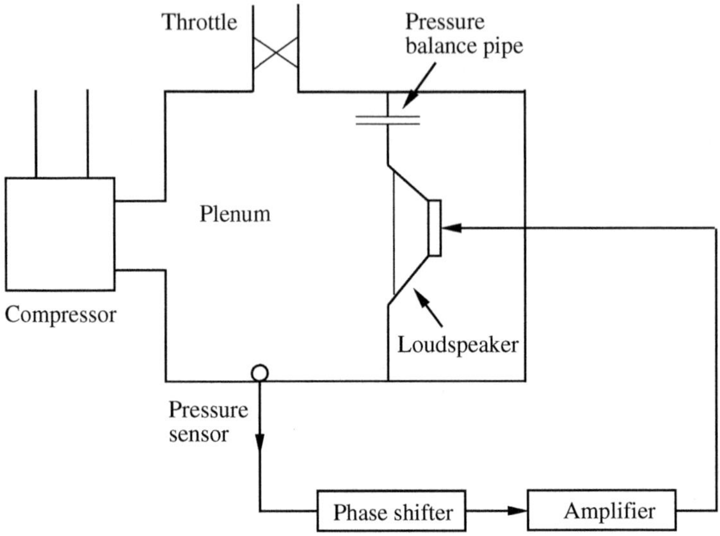

Fig. 9. Physical arrangement of an experimental compressor system in which surge was controlled using a feedback controller (after Hunang, 1987)

[18], in which a flow induced oscillation at 25 Hz was reduced by about 30dB. Although the loudspeaker drive voltage only needs to be small to suppress this oscillation provided the feedback system was working continuously as the car's speed increased, if control was lost for any reason the loudspeaker had to drive very hard for a few seconds to regain control.

A similar control system has been used by Hunang (1987) to control surge in a model compressor as shown in Fig. 9. Figure 10 shows the waveforms of the pressure and drive signal in this experiment, which have a similar form to those experienced in the control of sun roof oscillations described above, and the acoustic feedback control of flutter and the thermo-acoustic oscillation of a Rijka tube as reviewed, for example, by Ffowcs Williams (1996) [19] and Peak and Crighton (2000) [20].

The initial oscillations in the cavity and compressor described above can be described by simple second-order equations. It is tempting to speculate about the possibility of actively controlling much more complicated fluid phenomena which also generate noise, such as turbulence (Ffowcs Williams, 1996). There are many fundamental and practical problems in controlling such systems, partly because the order of the governing equations is large and the flow is three-dimensional, but some interesting work is being carried out in this field (Gad-el-Mak, 1998) [21].

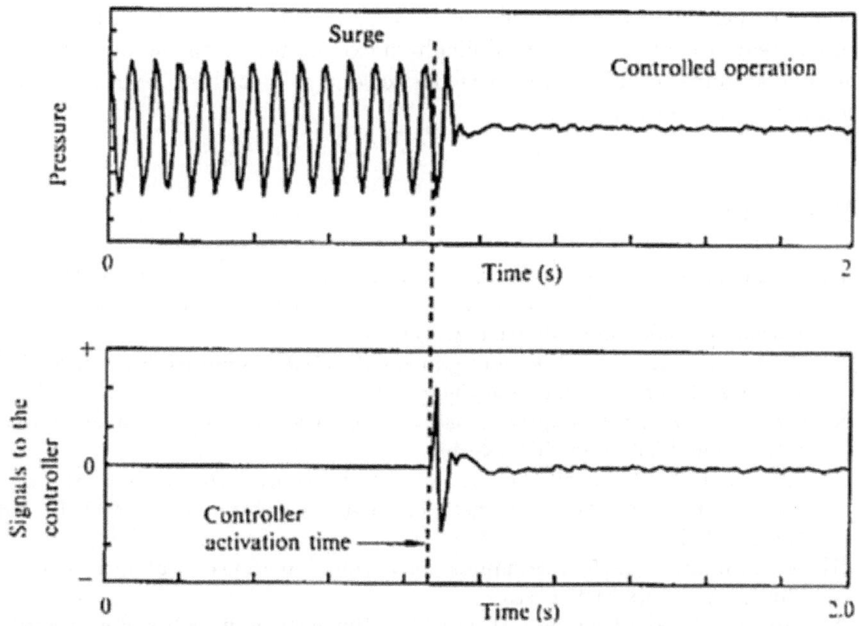

Fig. 10. Waveforms of the plenum pressure in the compressor and the signal to the control loudspeaker measured in the experimental control of surge by Hunang (1987)

7 Conclusions

The active control of low frequency sound propagating linearly in ducts and enclosures has been demonstrated by a number of authors. The physical mechanisms of control can be readily demonstrated in a one-dimensional duct in which a single secondary source can be used to reflect the sound wave and a pair of sources can be used to absorb it. Clear physical limitations exist on the upper frequency at which such control systems work, however, when limited to a reasonable number of control channels. nevertheless control systems with typically 40 actuators and 70 sensors are commercially fitted to control the low-frequency propeller noise inside aircraft (Ross and Purver, 1997) [23].

In many applications time-advanced information about the disturbance to be cancelled is available from an external reference signal, and so feedforward controllers are commonly used in active control. In order to maintain good performance in the face of non-stationary disturbances, practical control systems are almost always made adaptive and are thus closed loop.

In contrast to sound transmission, sound generation by aerodynamic sources is often a nonlinear phenomena, although the initial amplification of small disturbances can be approximated by linear models. The control of flow-induced pressure oscillations in cavities and surge in compressors is briefly discussed to illustrate how it may be possible for such systems to be suppressed at source. It is very difficult to extrapolate from these simple laboratory systems to many real aeroacoustic sources, but this does constitute a continuing area of research.

References

[1] Lueg P. (1936) Process of silencing sound oscillations. U.S. Patent, No.2,043,416
[2] Swinbank M.A. (1973) The active control of sound propagating in long ducts. Journal of Sound and Vibration 27:411–436.
[3] Clark R.L., Saunders W.R. and Gibbs G.P. (1998) Adaptive Structures, Dynamics and Control. John Wiley and Sons.
[4] Roure A. (1985) Self adaptive broadband active sound control system. Journal of Sound and Vibration 101:429–441.
[5] Elliott S.J. (2001) Signal Processing for Active control, Academic Press.
[6] Elliott S.J. (1998) Adaptive methods in active control. Proc. MOVIC '98, Zurich, 41–48.
[7] Fedorynk M.V. (1975) The suppression of sound in acoustic waveguides. Soviet Physics Acoustics 21:174–176.
[8] Zander A.C. and Hansen C.H. (1992) Active control of higher-order acoustic modes in ducts. Journal of the Acoustical Society of America 92:244–257.
[9] Burdisso R.A., Thomas R.H., Fuller C.R. and O'Brien W.F. (1993) Active control of radiated inlet noise from turbofan engines. Proc. Second Conference on Recent Advances in the Active Control of Sound and Vibration, 848–860.
[10] Joseph P., Nelson P.A. and Fisher M.J. (1996) An in-duct sensor array for the active control of sound radiated by circular flow ducts. Proc. InterNoise 96, 1035–1040.
[11] Nelson P.A. (2000) Active techniques and their potential for applications in aeroacoustics. Proc. 6th AIAA/CEAS Aeroacoustics Conference.
[12] Nelson P.A. and Elliott S.J. (1992) Active Control of Sound, Academic Press.
[13] Olson H.F. and May E.G. (1953). Electronic sound absorber. Journal of the Acoustical Society of America 25:1130–1136.
[14] Garcia-Bonito J. and Elliott S.J. (1996) Local active control of vibration in a diffuse bend wave field. ISVR Technical Memorandum No. 790.
[15] Elliott S.J., Joseph P., Bullmore A.J. and Nelson P.A. (1988) Active cancellation at a point in a pure tone diffuse field. Journal of Sound and Vibration 120:183–189.
[16] Fuller C.R., Elliott S.J. and Nelson P.A. (1996) Active Control of Vibration, Academic Press.
[17] Billout G., Galland M.A., Huu C.H. and Candel S. (1991) Adaptive control of instabilities. Proc. First Conference on Recent Advances in the Active Control of Sound and Vibration, 95–107.
[18] Stothers I.M., Saunders T.J., McDonald A.M. and Elliott S.J. (1993). Adaptive feedback control of sunroof flow oscillations. Proc. Institute of Acoustics 15(3):383–394.
[19] Ffowcs Williams J.E. (1996) Aeroacoustics. The Aeronautical Journal, December 1996, 531–537.

[20] Peak N. and Crighton D.G. (2000) Active Control of Sound. Annual Review of Fluid Mechanics 32:137–164.

[21] Gad-el-Hak M. (1998) Frontiers of flow control - Chap. 2 of Flow Control, Fundamentals and Practices. Gad-el-Hak M, Pollard A. and Bonnet J-P (Eds.) Springer.

[22] Sunyach M. and Ffowcs Williams J.E. (1986) The active suppression of oscillation in flow-excited cavities. C.R. Acad. Sc., Paris 303: II(12), 1085–1089.

[23] Ross C.F. and Purver M.R.J. (1997) Active cabin noise control. Proc. ACTIVE97, xxxix–xlvi.

Distributed Control and Observation

Claude Bardos

University Denis Diderot and LANN University Pierre et Marie Curie Paris
bardos@math.jussieu.fr

Abstract. Over the recent years, mostly under the impetus of the late JL Lions, important progress have been made for the control of distributed systems. This has contributed to the understanding of the duality which exist between the modal analysis and the need of very localized actuators. This duality leads to the phenomena of overspilling (excitation of higher order modes). On the other hand this type of research is closely related to the analysis of the exterior problem for the acoustic equation which was, under the influence of P. Lax, one of the main stimulus for the development of microlocal analysis. In this contribution only the acoustic equation is studied. However most of the idea carry on to other linear equations and to some non linear equations when considered in a perturbative regime where linearization techniques can be used; this includes perturbations of given solution of the Navier Stokes equation following the work of Imanuvilov and Fursikov (cf [12]) for a recent reference. Same ideas are used also in identification problems for instance in the migration method for oil recovery.

1 Introduction

In the classical theory of control for dynamical systems one considers the issues of controllability, observability, attainability and stabilization. The program of J.L. Lions was to systematically extrapolate these questions to partial differential equations. Now it turns out that, with the high frequency approximation, hyperbolic problems are the one that share the most common features with ordinary differential equation. Furthermore they have genuine application that stimulate theoretical progress.

One of these motivations was the control of the vibrations of large scale structures, like satellites antennas and the collaboration with mathematicians goes back also, to the best of my knowledge to a workshop organized with the NASA in Blacksburg 1984 where model problems were defined to help the understanding of the phenomena.

And more recently it was observed that the devices of time reversal, developed in the “Laboratoire Ondes et Acoustique” at the “Ecole de Physique Chimie de Paris” by Mathias Funk and his coworkers is the most perfect illustration of the the method presented in this note which focuses on the acoustic equation.

2 The Acoustic Equation

In an homogenous media Maxwell equations, linearized version of elasticity equations, and many others can be reduced to the equation

$$(\partial_t^2 - \Delta_x)u = 0 \,. \tag{1}$$

However there are several good reasons to consider in some cases a slightly more general form:

$$\partial_t^2 u - \frac{1}{\sqrt{\det g(x)}} \nabla_x (A(x)\sqrt{\det g(x)}\nabla_x u) = 0 \tag{2}$$

where Δ_x is replaced by the operator

$$u \mapsto \Delta_g u = \frac{1}{\sqrt{\det g(x)}} \nabla_x (A(x)\sqrt{\det g(x)}\nabla_x u) \,. \tag{3}$$

In (2) and (3) $A(x) = (a^{ij}(x))$ is a real $n \times n$ strictly positive symmetric matrix with smooth variable coefficients, $g(x) = A^{-1}(x)$ defines a riemannian metric with a volume element

$$dv(x) = \sqrt{\det g(x)}dx \,. \tag{4}$$

In fact several practical problems involve media which are no more homogeneous and therefore the use of space dependent coefficients and equation of the type (2) becomes compulsory. Furthermore details proofs, which may be omitted in the present contribution, require change of space variables which do transform the (1) into the (2).

We consider problems in a *bounded* domain Ω with boundary $\partial\Omega$ and exterior normal denoted by $\vec{n}(x)$ Standard such conditions are the Dirichlet boundary condition or the Neuman boundary condition:

$$u(x,t) = g(x,t) \quad \text{on } \partial\Omega \tag{5}$$

or

$$\frac{\partial u}{\partial n} = h(x,t) \quad \text{on } \partial\Omega \,. \tag{6}$$

The term homogenous is used to characterize situations where g or h are identically zero.

Finally to stabilize vibrating structures one may introduce devices (for instance piezo electric transducers in acoustic) which absorb energy (cf. (10) below) and which are described by an impedance boundary condition on a subdomain Γ_i of $\partial\Omega$

$$\partial_t u + Z(x)\frac{\partial u}{\partial n} = 0 \quad \text{on } \Gamma_i \quad \text{with } Z(x) \geq 0 \,. \tag{7}$$

The energy of the wave at time t is the sum of its kinetic and potential energy according to the formula:

$$E(t) = \frac{1}{2}\int \{|\partial_t u|^2 + |\nabla_x u|^2\} dx \,. \tag{8}$$

Multiplying the (1) by $\partial_t u$ and integrating by part over Ω one obtains the relation:

$$\frac{dE(t)}{dt} - \int_{\partial\Omega} \frac{\partial u}{\partial n}\partial_t u d\sigma_x = 0 \tag{9}$$

which shows that for the condition (5) and (6) the energy is an invariant of the motion and that for the condition (7) it decays according to the formula:

$$\frac{dE(t)}{dt} + \int_{\partial\Omega} Z(x)\left|\frac{\partial u}{\partial n}\right|^2 \partial_t u d\sigma_x = 0 \,. \tag{10}$$

It is convenient to introduce the unbounded operator $\mathcal{A}$ acting on the pair $U = (u(x,t), v(x,t)) = (u(x,t), \partial_t u(x,t))$ according to the formula

$$\mathcal{A}U \begin{pmatrix} v \\ \Delta_x u \end{pmatrix} = \begin{pmatrix} 0, & I \\ \Delta_x, & 0 \end{pmatrix} U \tag{11}$$

with a convenient definition of its domain $D(\mathcal{A})$ which takes in account the boundary condition. With this notation the wave equation is equivalent to the equation

$$\partial_t U = \mathcal{A}U \,. \tag{12}$$

The following very classical facts are consequence of functional analysis.

Theorem 1. *1. With homogenous Dirichlet or Neuman boundary condition the operator $\mathcal{A}$ is the generator of a group in the energy space*

$$\{H^1(\Omega), \quad u = 0 \quad \text{on} \quad \Gamma_D\} \times L^2(\Omega) \,.$$

with Γ_D denoting the part of the boundary where the homogenous Dirichlet boundary condition is assumed.

The solution of the homogenous Dirichlet or Neuman problem

$$(\partial_t^2 - \Delta_x)u = 0 \, u(x,0) \equiv 0, \quad \partial_t u(x,0) = f(x) \,. \tag{13}$$

can be expressed in term of the eigenfrequencies ω_k *and eigenmodes* ϕ_k *of the operator according to the formula:*

$$u(x,t) = \sum_{1\leq k<\infty} \frac{\sin(\omega_k t)(f,\phi_k)}{\omega_k}\phi_k \tag{14}$$

2. Assume that the boundary $\partial\Omega$ *is the union of three parts* Γ_D, Γ_N *and* Γ_I *then the operator* $\mathcal{A}$ *defined with homogenous Dirichlet boundary data on* Γ_D, *homogenous Neumann boundary data on* Γ_N *and impedance boundary condition on* Γ_I *is the generator is the generator of a contraction semi group in the space* $\{u \in H^1(\Omega),\ u = 0 \text{ on } \Gamma_D\} \times L^2(\Omega)$. *Furthermore if* Γ_I *is of positive measure in* $\partial\Omega$ *with* $Z(x) > 0\ \forall x \in \Gamma_I$ *then semi group* $e^{t\mathcal{A}}$ *converge strongly to zero with* $t \to \infty$. *Ie for any initial data* U *one has*

$$\lim_{t\to\infty} e^{t\mathcal{A}}U = 0\,. \tag{15}$$

The above statements are very classical but deserves the following comments.

1. The fact that the evolution of the solution is described by a group of operator is the functional analysis transcription of the conservation of the total energy (kinetic and potential)

$$\frac{\partial}{dt}\frac{1}{2}\int_\Omega \{|\partial_t u|^2 + |\nabla_x u|^2\}dx = 0\,.$$

Observe that for solutions of the Homogenous Dirichlet problem (and for T given) the following norms are equivalent

$$\frac{1}{2}\int_\Omega \{|\partial_t u(x,0)|^2 + |\nabla_x u(x,0)|^2\}dx, \int_{]0,T[\times\Omega} |\nabla_x u(x,0)|^2 dxdt$$

$$\text{and} \quad \int_{]0,T[\times\Omega} |\partial_t u(x,0)|^2 dxdt\,.$$

2. The formula (14) seems very appealing but its main drawback is the fact that in general the eigenfrequencies and eigenmodes are among the main unknowns of the problem. Furthermore each eigenvector is a solution of an elliptic equation:

$$-\Delta\phi_k = \omega_k^2\phi_k \tag{16}$$

and therefore cannot be localized (ie they cannot vanish on a set of positive measure).

3. The proofs of (15) follows (by a contradiction argument) from the fact that in the corresponding situations there exist no finite energy stationary solution of the problem. No information is given on the nature of the decay to zero.

However one could use the two followings facts. The domain of any power $\mathcal{A}^s$ $(0 \leq s)$ is conserved by the evolution and for $0 < s$ this domain is compactly injected in the space of solutions of finite energy. Therefore the Banach Steinhaus theorem implies the existence of function $f(t,s)$ with

$$\text{for } s > 0 \lim_{t \to \infty} f(t,s) = 0$$

such that one can supplement the relation (15) by the estimate.

$$\frac{1}{2} \int_{\Omega} \{|\partial_t u(x,t)|^2 + |\nabla_x u(x,t)|^2\} dx \leq f(t,s) \|(u(.,0), \partial_t u(.,0)\|_{D(\mathcal{A}^s)} \,. \quad (17)$$

Observe that such statement contains no information on the rate of decay of the function $f(t,s)$

The nature of the decay in the above problem can be explained in term of fronts and rays and it turns out that it is closely related to the question of observability which will by duality leads to the question of controllability. As a consequence further developments need some comments about the "sympletic geometry" related to the domain.

2.1 The Compressed Broken Hamiltonian Flow

2.1.1 The Hamiltonian Flow Inside Ω

To emphasize the effect of the inhomogeneities in the structure of the underling Hamiltonian flow in the following sections one considers the equation

$$\partial_t^2 u - \frac{1}{\sqrt{\det g(x)}} \nabla_x (A(x) \sqrt{\det g(x)} \nabla_x u) = 0 \quad (18)$$

Its principal symbol is the Hamiltonian

$$H(x,t,\xi,\tau) = \frac{1}{2} \{(A(x)\xi, \xi) - |\tau|^2\} \quad (19)$$

whose integral curves $\gamma(s) = (x(s), t(s), \xi(s), \tau(s)$ are the solutions of the differential system:

$$\frac{dx}{ds} = \partial_\xi H = A(x)\xi, \qquad \frac{dt}{ds} = \partial_\tau H = -\tau \quad (20)$$

$$\frac{d\xi}{ds} = -\partial_x H = -\frac{1}{2} \nabla_x \langle A(x)\xi, \xi \rangle \qquad \frac{d\tau}{ds} = 0 \quad (21)$$

Since both $H(x(s), t(s), \xi(s), \tau(s))$ and $\tau(s)$ (with the second equation of (21)) are invariants, the action of the flow can be reduced to the intersection of the wave cone with the spherical bundle ie

$$C = \{(x,\xi,t,\tau)\backslash(A(x)\xi,\xi) - |\tau|^2 = 0,\ |\tau| = 1\} \quad \text{or} \quad (A(x)\xi,\xi) = |\tau|^2 = 1 \tag{22}$$

and the mapping $\gamma(0) \mapsto \gamma(s)$ defines a foliation of C .

Denote by π the projection of these curves on $\mathbf{R}^n_x \times \mathbf{R}_t$ or (with the same notation) on $\mathbf{R}^n_x$:

$$\pi(\gamma(s)) = (x(s), t(s)) \tag{23}$$

and observe that the Hamiltonian $\frac{1}{2}(A(x)\xi,\xi)$ is the Legendre transform of the Lagrangian:

$$L(x,\dot{x}) = \frac{1}{2}(g(x)\dot{x},\dot{x}) \quad \text{with} \quad g(x) = A^{-1}(x)\,. \tag{24}$$

As a consequence the projection $x(s) = \pi\gamma(s)$ of the bicharacteristics are the geodesics ie the curves which make stationary the distance, or the energy, between two points for the riemannian metric defined by $g(x)$ (with volume element given by (4)) ie the quantities

$$\int_0^T \sqrt{\langle g(x)\dot{x},\dot{x}\rangle}\,dt \quad \text{and} \quad \int_0^T \frac{1}{2}\langle g(x)\dot{x},\dot{x}\rangle\,dt\,. \tag{25}$$

2.1.2 Interaction with the Boundary

Most of the information is carried by rays which propagate inside the domain, reflect on the boundary or are confined on this boundary. This set is called the *compressed broken hamiltonian flow* and can be defined as follow.

Starts with broken rays bicharacteristic $\gamma(s) \in \Omega$ which for $0 \le s < s^*$ satisfy the relation $x(s) \in \Omega$ and which for $s = s^*$ intersect the boundary transversally:

$$\langle \dot{x}(s^*), \vec{n}\rangle = \langle A(s^*)\xi(s^*), \vec{n})\rangle > 0 \tag{26}$$

with $\vec{n}$ denoting the normalized $((A(x)\vec{n},\vec{n}) = 1)$ outward normal to $\partial\Omega$. The reflected ray starts from the same point x at time $t(s^*)$, with initial speed $\dot{x}^{out}$:

$$\dot{x}^{out} = \dot{x}(s^*) - 2\langle \dot{x}(s^*)\cdot\vec{n}\rangle A(x)\vec{n} \tag{27}$$

With the equation $\dot{x} = A(x)\xi$ the reflection mapping:

$$(x(s^*), t(s^*), \dot{x}(s^*), \dot{t}(s^*)) \mapsto (x(s^*), t(s^*), \dot{x}^{out}(s^*), \dot{t}(s^*)) \tag{28}$$

defined on the tangent space is equivalent to the mapping

$$(x(s^*), t(s^*), \xi(s^*), \tau(s^*)) \mapsto (x(s^*), t(s^*), \xi^{out}(s^*), \tau(s^*)) \tag{29}$$

defined on the cotangent space with

$$\xi^{out} = \xi(s^*) - 2\langle A(x(s^*))\xi(s^*)\cdot\vec{n}\rangle\vec{n}\,. \tag{30}$$

By iteration this construction defines the broken flows $s \mapsto \gamma(s)$.

Then one introduce on the spherical bundle Ω the equivalence relation R defined, (for points $(x,\xi,t,\tau)\in\Omega$ with $x\in\partial\Omega$) by (30). This equivalence relation turns any broken flow into a C^0 map from $\mathbf{R}$ with value in Ω/R.

The *compressed broken hamiltonian flow* is defined by taking the C^0 closure of broken rays. It is contained in the unions of broken rays, grazing rays, and creeping rays and in non pathologic cases (contact of infinite order cf. Hormander [11]) example 24.3.11 page 438 volume III) it coincides with the union of such rays. In the sequel this is assumed under the *hypothesis of simple contacts.* In fact the details of this construction are involved and besides definitions requires technical proofs which are out of the scope of this presentation.

With the relations

$$\frac{dt}{ds}=-\tau \quad \frac{d\tau}{ds}=0 \tag{31}$$

in some cases it is convenient and adapted to the intuition to parametrized the ray (bicharacteristic) with the variable t

$$(x(s),t(s),\xi(s)\tau(s))=(x(t),t,\xi(t),\tau)\,. \tag{32}$$

Finally with the above construction one can define the geodesic distance between two points $(x,y)\in\Omega\times\Omega$ according to the formulas

$$d(x,y)=\inf\int_0^1|\dot{x}(s)|ds \quad \text{for} \quad x(s)\in C^1(\Omega), \quad x(0)=x, \quad x(1)=y \tag{33}$$

and the distance between a subset Γ and the open set Ω by the formula:

$$d(\Gamma,\Omega)=\sup_{y\in\Omega}\big(\inf_{x\in\Gamma}d(x,y)\big) \tag{34}$$

3 Control and Observation

The model control problem goes as follow: With the standard theorems on the mixt Cauchy problem:

$$\partial_t^2u_g-\Delta u_g=0 \text{ in } \Omega\times]0,T[,\, u_g(x,0)=\partial_t u_g(x,0)=0 \tag{35}$$

$$u_g(x,t)=g(x,t) \text{ sur } \partial\Omega\times]0,T[,\ \text{supp } g\subset\Gamma\times]0,T[\tag{36}$$

defines a mapping

$$g\mapsto C(g)=(u_g(.,T),\partial_t u(.,T)) \tag{37}$$

from $L^2(\Gamma\times]0,T[)$ with value in $L^2(\Omega)\times H^{-1}(\Omega)$.

Address the following questions:

1 Is the set $\Gamma\times]0,T[$ large enough to ensure that the mapping $g\mapsto C(g)$ is of dense range in this case In this case the term *approximate controllability* is used.

2. Is the set $\Gamma\times]0,T[$ large enough to ensure that the mapping $g \mapsto C(g)$ is onto in this case one would say that *exact controllability holds.* By linearity the exact controllability is in the present case equivalent to the so called nul controllability:

2′. Is the set $\Gamma\times]0,T[$ large enough to ensure for any $(u_0(.,0)\,,\,u_1(.,0)) \in L^2(\Omega)\times H^{-1}(\Omega)$ the existence of a function $g\in L^2(\Gamma\times]0,T[)$ which brings to rest the solution after the time T ie such, that for the problem

$$\partial_t^2 u_g - \Delta u_g = 0 \quad \text{in} \quad \Omega\times]0,T[, u_g(x.0) = u_0(x)\,, \partial_t u_g(x,0) = u_1(x)\,, \tag{38}$$

$$u_g(x,t) = g(x,t) \quad \text{on} \quad \partial\Omega\times]0,T[,\ \text{supp}\ g \subset \Gamma\times]0,T[\tag{39}$$

one has:

$$u_g(x,T) = \partial_t u_g(x,T) = 0 \quad \text{on} \quad \Omega\,. \tag{40}$$

By a simple duality argument (using the fact that the problem is linear and time reversible) Lions has shown under the name of HUM method that the problem 1 and 2 (concerning the size $\Gamma\times]0,T[$ where equivalent to the following *observation* and *stable observation* property.

Observation Observation from $\Gamma\times]0,T[$ is possible if this set is large enough to ensure the following property: For any solution of

$$\partial_t^2 u - \nabla(A(x)\nabla u) = 0 \quad \text{in} \quad \Omega\times]0,T[, u(x,t) = 0 \quad \text{on} \quad \partial\Omega\times]0,T[\tag{41}$$

one has the implication:

$$\partial_{\vec{n}} u = 0 \quad \text{on} \quad \Gamma\times]0,T[\Rightarrow u \equiv 0\,. \tag{42}$$

Stable observation The observation is stable if there exits a constant $C<\infty$ such that for any solution of (41) one has:

$$E(u) \le C\int_0^T\int_\Gamma |\partial_n u|^2 d\sigma dt\,. \tag{43}$$

It is clear, and will become even more obvious with geometric considerations that the statement (43) is much stronger than the statement (42). Furthermore a simple argument [4] shows that if there exists a time T which makes $\Gamma_i\times]0,T[$ large enough to ensure, for the homogenous Dirichlet problem the estimate

$$E(u) \le C\int_0^T\int_{\Gamma_i} |\partial_n u|^2 d\sigma dt \tag{44}$$

then the decay in the stabilization formula (17) is uniform and more precisely one has

$$\int_\Omega \{|\partial_t u(x,t)|^2 + |\nabla_x u(x,t)|^2\}dx \le Ce^{-2\beta t}\int_\Omega \{|\partial_t u(x,0)|^2 + |\nabla_x u(x,0)|^2\}dx\,. \tag{45}$$

To motivate the precise statements which will be given in the next section one can make the following

Remark 1. It is known since Huyghens that the solution of the wave equation propagate with finite speed therefore to observe any signal which propagates in Ω one has to wait long enough to ensure that any signal generated in Ω at time $t = 0$ will reach Γ before the time T. This implies, as a necessary condition, for the observation, the relation (with the geodesic distance) :

$$T > 2d(\Gamma, \Omega) = 2 \sup_{y \in \Omega} \left(\inf_{x \in \Gamma} d(x, y) \right) . \tag{46}$$

To identify all the details of the phenomena with an arbitrary large precision one has to consider its high frequency components which propagate like rays therefore one has to "catch" all the rays.

At variance if the signal is "hidden" due, for instance, to the geometry of the domain then only the diffracted part or the front will be observed. In this situation the information will not be meaningful for wave with highly oscillatory components.

Finally if the set $\Gamma \times]0, T[$ is large enough to ensure the observation, with the Banach Steinhaus theorem one can show (cf (17)) the existence of a function $f_s(\alpha, \beta)$ with the following properties: For all solution u of (41) one has:

$$\begin{aligned} &\forall s > 0 \|u\|_{L^2(\omega \times]0,T[)} \leq f_s \\ &\times \left(\int_0^T \int_\Gamma |\partial_n u|^2 d\sigma dt, \|u\|_{H^s(\omega \times]0,T[)} \right), \lim_{\alpha \to 0} f_s(\alpha, \beta) = 0 . \end{aligned} \tag{47}$$

Remark 2. The relation (47) can be interpreted as the analysis of a feed back process. The action is $\partial_{\vec{n}} u$. It is computed in term of the observation $\partial_t u$ through the impedance relation:

$$\partial_n ou = -\frac{1}{Z(x)} \partial_t u$$

and (feedback) claims that in all cases the feed back has some efficiency if the initial wave does not oscillate to much.

4 Observation Estimates

4.1 Unstable Observations Estimates

In fact it was observed by Holmgren in 1901 [10] that for analytic coefficients the condition (46) is also sufficient for the observation. Such a result was improved by John [13] (he obtained some type of estimates concerning the conditional stability) and finally Lebeau [15] using the method of Leray [17] obtained the following estimate.

$$||u||^2_{L^2} \leq \frac{CE(u)}{\left[\log^+\left(\frac{E(u)}{\int_{\Gamma\times]0,T[}|u(\sigma,t)|^2 d\sigma dt}\right)\right]^2} . \qquad (48)$$

Proofs rely on sharp a priori estimates for the solution of the Cauchy Kowalewsy problem and duality methods.

Remark 3. 1) With the existence of gaussian beams concentrated, as this will be evoked in the next section, up to an exponentially small error term (when the matrix $A(x)$ is analytic) near a ray which does not meet the region $\Gamma\times]0,T[$ the above result is in a certain sense optimal.

2) (48) can be view as a low frequency estimate because it has a physical meaning only when the quantity

$$\frac{E(u)}{||u||^2_{L^2}}$$

is not too large and in the absence of any explicit Fourier decomposition this quantity is the most natural candidate to measure the frequency of the wave.

The analyticity or at least (cf [22]) the fact that the wave operator is independent of t seems to play a crucial role. It was observed by Alhinac and Baouendi [1] that one can construct in $\mathbf{R}^n \times \mathbf{R}_t$ a wave equation with C^∞ zero order time dependent coefficients

$$\partial_t^2 u - \Delta u + q(x,t)u = 0$$

which has at least one non zero solution wich is identically equal to zero in the halfspace $x_1 < 0$. In this case there is no unique continuation principle and no relation of type (48). However for time independent coefficients the unique continuation property holds and the problem is partly solved due to a serie of contributions ([22] and [20]). Same estimates have been obtained [16] for the problem of stabilization and giving for (17) the relation:

$$\frac{1}{2}\int_\Omega \{|\partial_t u(x,t)|^2 + |\nabla_x u(x,t)|^2\}dx \leq \frac{C}{\log^+ t}\|(u(.,0),\partial_t u(.,0)\|_{D(A^s)} . \qquad (49)$$

The main tools are the Carleman estimates and pseudo differential calculus with parameter. In fact this type of results are the most convincing applications of Carleman estimates and pseudodifferential calculus with parameter for the wave operator

$$P = \partial^2_{tt}\cdot - \nabla_x \cdot (A(x)\nabla_x \cdot) .$$

It involves computations of quantities of the form:

$$\int_{-\infty}^{\infty}\int_{\mathbf{R}^3}(e^{-k\phi}P(e^{k\phi})v)\,,\,v)dxdt$$

where v denotes a function with compact support in space time. The constant k will be taken large enough. In the final estimates k it appears with a factor $-l$ in the estimation of the derivative of order l of v and therefore it give the same weight to low and high frequency. The function ϕ satisfy the relation:

$$|\partial_t\phi|^2 - (A(x)\nabla_x\phi, \nabla_x\phi) \neq 0 \tag{50}$$

on the support of v it is strictly convex with respect to the propagation of the "fronts" and therefore seems to emphasize the role of the fronts compared to the role of the rays.

4.2 Propagation Along Rays and High Frequency Estimates

This section uses the high frequency asymptotic and therefore it is in full agreement with intuition leading to necessary and almost equivalent sufficient conditions. These conditions are very restrictive and this is in full agreement with the fact that more localized are the actuators bigger is the chance of the "spillover" or the excitation of higher order modes and this is in agreement with the remark (2).

The basic theorem of propagation of singularities (with the definition of Wave front set given by Hormander) is

Theorem 2. *For any compressed broken ray γ there exists a solution u of the corresponding wave equation*

$$\partial_t^2 u - \frac{1}{\sqrt{detg(x)}}\nabla_x(A(x)\sqrt{detg(x)}\nabla_x u) = 0, \text{ in } \Omega\times\mathbf{R}_t \; u(x,t) = 0 \; \forall x \in \partial\Omega \tag{51}$$

with $WF(u) = \gamma$.

From this statement one deduces the following

Proposition 1. *1. Consider the wave equation in a bounded domain $\Omega\times]0,T[$ with Dirichlet boundary condition on $\partial\Omega\times]0,T[$*

$$(\partial_t^2 - \Delta_x)u = 0 \quad u = 0 \quad \text{on} \quad \partial\Omega\times]0,T[\tag{52}$$

assume that there is at least one compressed broken ray which does not meet the closure (in $\partial\Omega\times\mathbf{R}_t$) of the observation region then there is no constant C such that for every solution holds the relation:

$$E(u) \leq C\int_0^T\int_\Gamma |\partial_n u|^2 d\sigma dt \tag{53}$$

and the exact controllability (from the same control region) cannot be achieved.

2. Consider the wave equation with an impedance boundary condition on Γ_i some part of the boundary of a bounded domain Ω .

$$(\partial_t^2 - \Delta_x)u = 0, \quad \partial_t u + Z(x)\partial_{\vec{n}} u = 0 \quad \text{on} \quad \Gamma_i, \quad u = 0 \quad \text{in} \quad \partial\Omega\backslash\Gamma_i \quad (54)$$

and denote by $e^{t\mathcal{A}}$ the associated semi group. Assume that Γ_i is not trapping ie. For any time $0 < T < \infty$ there exists a generalized bicharacteristic $(x(t), t, \xi(t), \tau)$ such that for $0 \le t \le T$

$$\{(x(t), t)\} \cap \overline{\Gamma_i \times [0, T]}\} = \emptyset \qquad (55)$$

then there exists no function $f(t)$ such that:

$$\|e^{t\mathcal{A}}U\|_{energy\ norm} \le f(t)\|U\|_{energy\ norm} \text{ and } \lim_{t\to 0} f(t) = 0\,. \qquad (56)$$

Remark 4. When the coefficients are constant or analytic, using the symplectic properties of the hamiltonian flow, one can construct (cf. [19]) solutions of "frequency" ω , $u_\omega(x,t)$ which for $\omega \to \infty$ concentrate exponentially along the ray $\gamma(s)$

$$|u(x,t) - u(x(s), s)| \le e^{-C(|x-x(s)|^2+|t-s|^2}\,. \qquad (57)$$

These solutions are called gaussian beams and their existence shows that the rates given by (48) and (49) are optimal.

With the definition of a diffractive point (ie a point where the ray is tangent to the boundary while remaining in $\overline{\Omega}$) cf [4] The proposition (1) can be complemented by the following

Theorem 3. *1 Consider the wave equation in a bounded domain $\Omega\times]0,T[$ with Dirichlet boundary condition on $\partial\Omega\times]0,T[$*

$$(\partial_t^2 - \Delta_x)u = 0 \quad u = 0 \quad \text{on} \quad \partial\Omega\times]0,T[\qquad (58)$$

and assume that all rays meet the observation region $\Gamma\times]0,T[$ in a non diffractive point, then there is a constant C such that for every solution holds the relation:

$$E(u) \le C\int_0^T\int_\Gamma |\partial_n u|^2 d\sigma dt \qquad (59)$$

and the exact controllability (from the same control) region can be achieved.

2 Consider the wave equation

$$(\partial_t^2 - \Delta_x)u = 0, \quad \partial_t u + Z(x)\partial_{\vec{n}} u = 0 \quad \text{on} \quad \Gamma_i, \quad u = 0 \quad \text{in} \quad \partial\Omega\backslash\Gamma_i \quad (60)$$

with an impedance boundary condition on Γ_i some part of the boundary of a bounded domain Ω and homogenous (Dirichlet or Neumann) on the rest of the boundary. Assume that there exists a time T such that all rays meet the impedance region $\Gamma_i\times]0,T[$ in a non diffractive point then there exist two constants C and $\beta > 0$ such that one has

$$\|e^{t\mathcal{A}}\|_{energy\ norm} \le Ce^{-\beta t}\,. \qquad (61)$$

The original proof of these theorems used a microlocal version of the propagation of regularity in Sobolev spaces as introduced by Melrose and Sjostrand [18]. Since it is the propagation of energy which is concerned it turns out that in some cases a much simpler proof with the introduction of defect measures is available. This is well documented in the expository talk of Burq [6].

5 Time Reversal Methods and Ergodicity

The material described above above turn out to be particularly relevant for the mathematical analysis of the so called time reversal method.

Basically proceed as follow. Observe a wave reflected by a media during a finite time $0 < t < T$ and send back the signal after time symetry with respect to T and may be some amplification. Use the invariance of the wave equation (with respect to the change $t \to 2T - t$) to reconstruct (up to an amplification factor and may be some error) the initial signal. In general this does not produces exact controllability but it replaces complicated numerical algorithms by real time action. Such devices are used in many applications ranging from oil recovery, non destructive control and so on ... to focalization of ultrasonic beams on kidneys stones. Since the media are very often ergodic or random a special attention has been given to this situation. It has been found out that instead of deteriorating the process randomness (cf. [2] for the most recent contribution in the field) or ergodicity [3] contribute to its success. The effect of the ergodicity has been experimented (real experiments and numerical simulations) by M. Fink and his group. The mathematical analysis of such experiment rely on recent theorems on "Qantum chaos" which show what can be achieved when the action is done on an arbitrary small domain but for large time. This is the subject of the next lines.

Consider the in a bounded domain Ω the solution of the initial value problem:

$$\partial_t^2 u_i - \Delta u_i = 0, \quad \text{in} \quad \Omega \quad u(x,t) \equiv 0 \quad \text{on} \quad \partial\Omega \tag{62}$$

with the initial condition:

$$u_i(x,0) = 0 \ \text{ and } \ \partial_t u(x,0) = \psi(x) \tag{63}$$

Observe for $0 < t < T$ as in the impedance problem the time derivative of this incident wave the solution $\partial_t u_i$ of (62) and (63) on a subset $\sigma \subset \Omega$ (which may be *very small*) introduces an L^∞ function $\Xi(x)$ with support contained in σ and eventually introduces for $T < t < 2T$ the solution of the problem:

$$\partial_t^2 u_r - \Delta u_r = \Xi(x)\partial_t u_i(x, 2T - t), \quad \text{in} \quad \Omega \quad u(x,t) = 0 \quad \text{on} \quad \partial\Omega \tag{64}$$

with initial conditions:

$$u_r(x,T) = u_i(x,T), \quad \partial_t u_r(x,T) = \partial_t u_i(x,T)\,. \tag{65}$$

Therefore one has:

$$\partial_t u_r(x,2T) = \cos\left(2T(-\Delta)^{\frac{1}{2}}\right)\psi + \int_T^{2T} \cos\left((2T-t)\,(-\Delta)^{\frac{1}{2}}\right) \Xi\cos\left((2T-t)(-\Delta)^{\frac{1}{2}}\right)\psi dt\,. \tag{66}$$

To use the ergodicity property T will be taken *large enough.* This also reinforces the influence of the reemitted signal. According to these observations one writes:

$$u_r(x,2T) = T\left\{\frac{1}{T}\cos(2T(-\Delta)^{\frac{1}{2}})\psi + \frac{1}{T}\int_T^{2T} \cos\left((2T-t)(-\Delta)^{\frac{1}{2}}\right) \Xi\cos\left((2T-t)(-\Delta)^{\frac{1}{2}}\right)\psi dt\right\} \tag{67}$$

In any convenient sense and in particular for the energy norm (with initial data of finite energy) one has

$$\lim_{T\to\infty}\frac{1}{T}\cos(2T(-\Delta)^{\frac{1}{2}})\psi = 0\,. \tag{68}$$

Therefore

$$u_r(x,2T) \simeq T\lim_{T\to\infty}\frac{1}{T}\int_T^{2T}\cos((2T-t)(-\Delta)^{\frac{1}{2}})\Xi\cos((2T-t)(-\Delta)^{\frac{1}{2}})\psi dt \tag{69}$$

whenever this limit exists.

One has

$$\begin{aligned}\frac{1}{T}\int_T^{2T} &\cos\left((2T-t)(-\Delta)^{\frac{1}{2}}\right)\ \Xi\cos\left((2T-t)(-\Delta)^{\frac{1}{2}}\right)\psi dt\\ &= \frac{1}{T}\left(\int_0^T \cos(t(-\Delta)^{\frac{1}{2}}\right)\Xi\cos\left(t(-\Delta)^{\frac{1}{2}})dt\right)\psi\\ &= \frac{1}{4T}\left(\int_0^T (e^{it(-\Delta)^{\frac{1}{2}}} + e^{-it(-\Delta)^{\frac{1}{2}}}\right)\\ &\quad \Xi\left(e^{it(-\Delta)^{\frac{1}{2}}} + e^{-it(-\Delta)^{\frac{1}{2}}})dt\right)\psi\,.\end{aligned} \tag{70}$$

The last term of (70) can be written as the sum of two terms:

$$M(T) = \frac{1}{4T}\left(\int_{-T}^T (e^{it(-\Delta)^{\frac{1}{2}}}\Xi\left(e^{it(-\Delta)^{\frac{1}{2}}}\right)dt\right)\psi$$

and

$$N(T) = \frac{1}{4T}\left(\int_{-T}^T (e^{it(-\Delta)^{\frac{1}{2}}}\Xi\left(e^{-it(\Delta)^{\frac{1}{2}}}\right)dt\right)\psi\,.$$

The term $M(T)$ contains no resonnancy and therefore with the Weil formula for the asymptotic behavior of the eigenfrequency one has the

Proposition 2. *The operator $M(T)$ converges weakly to zero with $T \to \infty$. More precisely for any pair of smooth functions with compact support $(\psi, \theta) \in (\mathcal{D}(\Omega))^2$ one has:*

$$|(\theta, M(T)\psi)| \leq \frac{C(\theta, \psi)}{T} \,. \tag{71}$$

Furthermore in dimension $d = 2$ the estimate (71) holds for the Hilbert Schmidt norm of $M(T)$

$$\|M(T)\|_{HS} \leq \frac{C}{T} \,. \tag{72}$$

Therefore the large time asymptotic behavior of the "reversed" wave is given by:

$$u_R(x, 2T) \simeq TN(T) \simeq \frac{T}{2}\left(\frac{1}{2T}\int_{-T}^{T}(e^{it(-\Delta)^{\frac{1}{2}}}\,\Xi\left(e^{-it(\Delta)^{\frac{1}{2}}}\right)dt\right)\psi \,. \tag{73}$$

To take in account ergodic properties of the hamiltonian flow one proceeds as follow:

The action of the the broken hamiltonian flow is extended as an operator on the functions defined on the cotangent spherical bundle $S^*(\bar{\Omega}) = \{(x, \xi)\ x \in \Omega, |\xi| = 1\}$ with volume denoted:

$$|S^*(\bar{\Omega})| = \iint_{\Omega \times \{|\xi|=1\}} dx d\xi \,.$$

The principal symbol of any zero order pseudo differential operator P is denoted by σ_P.

Definition 1. *(1) The flow is said to be classically ergodic if for any continuous function $f \in C^0(S^*(\bar{\Omega}))$ one has (in the weak L^* topology)*

$$\lim_{t\to\infty} V(t)f = \bar{f} = \frac{1}{|S^*(\bar{\Omega})|}\int_{\Omega \times \{|\xi|=1\}} f(x, \xi) dx d\xi \,. \tag{74}$$

(2) With the introduction of the projection Π_l on the space spanned by the l first eigenvectors of $-\Delta$

$$\Pi_l = \sum_{1\leq k\leq l} \phi_k \otimes \phi_k^* \tag{75}$$

an operator $K \in \mathcal{L}(L^2(\Omega))$ is said to be spectrally regularizing ($K \in SR$) if it satisfies the following estimate:

$$\|\Pi_l K \Pi_l\|_{HS}^2 = o(l), \tag{76}$$

– *(3) The flow is said to be quantum ergodic if, for any zero order pseudo differential operator P and in the weak operator limit, one has*

$$\lim_{T\to\infty}\frac{1}{2T}\int_{-T}^{T} e^{it(-\Delta)^{\frac{1}{2}}} P e^{-it(-\Delta)^{\frac{1}{2}}}dt = \langle P\rangle I + K \tag{77}$$

with $K\in SR$ and

$$\langle P\rangle = \frac{1}{|S^*(\bar\Omega)|}\int_{\Omega\times\{|\xi|=1\}}\sigma_P(x,\xi)dxd\xi = \lim_{l\to\infty}\sum_{1\le k\le l}\frac{1}{l}(P\phi_k,\phi_k)\,. \tag{78}$$

It has been proven that classical ergodicity implies quantum ergodicity (cf. [7, 21, 23] and ([24])

Therefore with the proposition 5.1 one has the

Theorem 4. *For the solution constructed according to the formulas (62), (63) and (64), (65) one has for $T\to\infty$ and in the sense of distributions the following asymptotic:*

$$u_R(x,2T)\simeq\frac{T}{2}((\langle\Xi\rangle(\psi+K\psi) \tag{79}$$

with $K\in SR$ and

$$\langle\Xi\rangle = \lim_{l\to\infty}\sum_{1\le k\le l}\frac{1}{l}(\Xi\phi_k,\phi_k)\,. \tag{80}$$

Remark 5. The notion of spectrally regularizing ($K\in SR$) is not in its present form very explicit. However for any pseudo differential operator P of zero order one has (cf. [25]) proposition 1.1 (ii))

$$\lim_{l\to\infty}\frac{1}{l}\|\Pi_l P\Pi_l\|^2_{HS} = \langle P\rangle = \frac{1}{|S^*(\bar\Omega)|}\int_{\Omega\times\{|\xi|=1\}}|\sigma_P(x,\xi)|^2 dxd\xi \tag{81}$$

therefore any pseudo differential operator P which belongs to SR has his principal symbol equal to zero and has a regularizing effect. Similarly one shows (cf [23]) page 921 that in general (at least when the spectra of Δ has bounded multiplicity) K is compact. This is the reason why the above theorem carries pertinent information when the initial data ψ is a distribution with a unique singularity located at one point says A then the reversed solution is a sum of a more regular term $\frac{T}{2}K\psi$ and of a leading term proportional to $\frac{T}{2}\psi$:

$$\frac{T}{2}\langle\Xi\rangle\psi\,. \tag{82}$$

As a conclusion the ergodicity property (ie the fact that for T large enough almost any ray meets the control region (support of Ξ) even if this support is very small (just wait long enough)) allows to reconstruct (with an amplification) a function that will have the same singularities as the initial data. If the initial data is very localized (say a δ function at a point A) then at the time $2T$ the signal will be a $T/2\delta$ at the same point A plus a smoother term which appears as a modulation. In this sense for large time the refocalization is perfect and this is in agreement with experiment and numerical simulations. [8] and [9].

6 Conclusion

This contribution is devoted to the extension of the issues raised by control theory for systems of ordinary differential equations to distributed systems described by partial differential equations. It has been shown that a systematic approach involves a frequency analysis of the solution. One uses the fact that solutions of linear hyperbolic problems propagate along rays. However due to the Heisenberg uncertainty principle this statement is true only in an asymptotically with the introduction of high frequency analysis (which in the present time carries the name of microlocal analysis). In the absence of the observation of all the high frequencies one can use the diffractive aspect of the wave to obtain positive results for low frequency regime. One can also balance the absence of total observation by long time observation when the underlying hamiltonian flow is ergodic.

References

[1] S. Alinhac and M. S. Baouendi, A non uniqueness result for operators of principal type, Math. Zeit. **220** (1995), No. 4, 561–568.

[2] G. Bal and L. Ryzhnik: Time reversal for classicla waves in random media *C. R. Acad. Sci. Paris t.* 332 (2001).

[3] C. Bardos and M. Fink: Mathematical foundations of the time reversal mirror *to appear in Asymptotic Analysis.*

[4] C. Bardos, G. Lebeau, and J. Rauch, Sharp sufficient conditions for the observation, Control and stabilization of waves from the boundary, *SIAM Journal on Control Theory and Applications*, **30** (1992), 1024–1065.

[5] N. Burq: Décroissance de l'énergie locale de l'équation des ondes pour le problème extérieur et absence de résonance au voisinage du réel, *Acta Math.*, **180**, (1988) 1–29.

[6] N. Burq: Mesures de default *Séminaire Bourbaki 1996–97 n 826 March 1997.*

[7] Y. Colin de Verdière: Ergodicité et fonctions propres du Laplacien. *Comm. Math. Phys*, **102**, (1985) 497–502.

[8] C. Draeger, M. Fink: One-channel time-reversal in chaotic cavities : Theoretical limits, *Journal of Acoustical Society of America*, **105** , 2, (1999), 611–617.

[9] C. Draeger and M. Fink: One-Channel Time Reversal of Elastic Waves in a Chaotic 2D-Silicon Cavity. *Phys. Rew. Letters*, **79**, 3, (1997) 407–410.

[10] E. Holmgren: Uber Systeme von linearen partiellen Differentialgleichnungen. *Ofversigt af Kongl. Vetenskaps-Akad. Forh*, **58**, (1901) 91–105.

[11] Hörmander L., The analysis of linear partial differential operators I, II, III, IV, Grundlehren der Math.Wiss. 256,257,274,275, Springer-Verlag (1985).

[12] O. Y. Imanuvilov: Remarks on exact controllability for the Navier-Stokes equations, *ESAIM Control Optim. Calc. Var.* **6** (2001), 39–72 (electronic).

[13] F. John, On linear partial differential equation with analytic coefficients, *Commun. Pure Appl. Math.* **2** (1949), 209–253.

[14] P. Lax and R. Phillips: Scattering Theory, *Academic Press (1989).*

[15] G. Lebeau: Contrôle analytique 1: estimations à priori, Duke Math. J. **68** (1992), 1–30.

[16] G. Lebeau et L. Robbiano: Stabilisation de l'équation des ondes par le bord, Duke Math. J. **86**, No. 3 (1997) 465–491.
[17] J. Leray: Uniformisation de la solution du problème linéaire analytique de Cauchy près de la variété qui porte les données de Cauchy (problème de Cauchy I), *Bull. Soc. Math. France* 85 (1957) 389–429.
[18] R. Melrose and J. Sjöstrand, Singularities of boundary value problems, *Comm. on Pure and Appl. Math.* **31**, 593 (1978).
[19] J.V. Ralston, Solutions of the wave equation with localized energy, Commun. on Pure and Appl. Math., **22** (1969), 807–823.
[20] L. Robbiano and C. Zuily Robbiano: Uniqueness in the Cauchy problem for operators with partially holomorphic coefficients. *Invent. Math.* 131 (1998), no. 3, 493–539.
[21] A.I Schnirelman: Ergodic properties of eigenfunctions. *Usp. Math. Nauk*, **29**, (1994) 181–182.
[22] D. Tataru: Unique continuation for operators with partially analytic coefficients. *J. Math. Pures Appl.* (9) 78 (1999), no. 5, 505–521.
[23] S. Zelditch., Uniform distribution of eigenfunctions on compact hyperbolic surfaces. *Duke Math. J.* **55**, (1987) 919–941.
[24] S. Zelditch., Quantum transition amplitudes for ergodic and for completely integrable systems. *J. Funct. Anal.*, **94**, (1990) 415–436.
[25] S. Zelditch: Quantum mixing *J. Funct. Anal.*, **140**, (1996) 68–86.

Global Boundary Stabilization of 2D Poiseuille Flow*

Andras Balogh and Miroslav Krstic†

Department of MAE, University of California at San Diego, La Jolla
CA 92093-0411

1 Introduction

In this article we address the problem of boundary control of a viscous incompressible fluid flow in a 2D channel. Many advances have been made on this topic by Bewley and coworkers [4], Speyer and coworkers [8, 20, 21], and others employing linear optimal control techniques in the CFD setting. Much progress has also been made on the topic of *controllability* of Navier–Stokes equations, a prerequisite to stabilization and optimal control problems.

Our objective here is to globally *stabilize* the parabolic equilibrium profile in channel flow. This objective is different than the efforts on optimal control [2, 9, 10, 13, 15, 16, 18, 27] or controllability [6, 11, 12, 14, 17] of Navier–Stokes equations. Optimal control of nonlinear equations such as Navier–Stokes is not solvable in closed form, forcing the designer to either linearize or use computationally expensive finite–horizon model–predictive methods. Controllability-based solutions, while a prerequisite to all other problems, are not robust to changes in the initial data and model inaccuracies. The stabilization objective indirectly addresses the problems of turbulence and drag reduction, which are explicit in optimal control or controllability studies. Coron's [7] result on stabilization of Euler's equations is the first result that directly addresses flow stabilization.

The boundary feedback control we derive in this paper is fundamentally different from those in [4, 8, 20, 21], which use *wall normal* blowing and suction. Our analysis motivated by Lyapunov stabilization results in *tangential* velocity actuation. Tangential actuation is technologically feasible. The work on synthetic jets of Glezer [26] shows that a teamed up pair of synthetic jets can achieve an angle of 85° from the normal direction with the same momentum as wall normal actuation. The patent of Keefe [22] provides the

*This work was supported by grants from the Air Force Office of Scientific Research, the National Science Foundation and the Office of Naval Research.

†Author to whom correspondence should be addressed

means for generating tangential velocity actuation using arrays of rotating disks.

An implementational advantage in our result is that, while it uses only the measurement of wall shear stress as in the previous efforts, it employs it in a *decentralized* fashion. This means that the feedback law can be embedded into the MEMS hardware (without need for wiring).

While it is possible and important to achieve higher order $(H^1$ and $H^2)$ stability, due to space limitations we restrict the analysis to the L^2 case in the present paper. For more details the interested reader is referred to [1].

The only limitation in our result is that it is guaranteed only for sufficiently low values of the Reynolds number. In simulations we demonstrate that the control law has a stabilizing effect far beyond the value required in the theorem (five or more orders of magnitude).

The paper is organized as follows. We formulate our problem in Sect. 2 and design boundary feedback laws in Sect. 3. Our main results are stated in Sect. 4 and L^2 stability is proved in Sect. 5 by employing Lyapunov techniques. In Sect. 6 we give numerical demonstrations that strengthen our theoretical results. Finally, in Sect. 7, we extend our numerical simulation to the case when the control is restricted to only part of the channel wall.

2 Problem Statement

The channel flow can be described by the 2D Navier–Stokes equations

$$\begin{cases} \mathbf{W}_t - \nu \Delta \mathbf{W} + (\mathbf{W} \cdot \nabla)\, \mathbf{W} + \nabla P = 0\,,\, 0 < x < 1,\, 0 < y < \ell,\, t > 0\,, \\ \qquad\qquad\qquad\qquad \mathrm{div}\mathbf{W} = 0\,,\, 0 < x < 1,\, 0 < y < \ell,\, t > 0\,, \end{cases} \tag{1}$$

where $\mathbf{W} = \mathbf{W}(x,y,t) = (U(x,y,t), V(x,y,t))^T$ represents the velocity vector of a particle at (x,y) and at time t, $P = P(x,y,t)$ is the pressure at (x,y) and at time t, $\nu > 0$ is the kinematic viscosity and the positive constant l represents the width of the channel. Our goal is to regulate the flow to the parabolic equilibrium profile (see Fig. 1)

$$\overline{U}(y) = \frac{a}{2\nu} y (\ell - y)\ , \tag{2}$$

$$\overline{V} = 0\,, \tag{3}$$

$$\overline{P}(x) = -ax + b\,, \tag{4}$$

where $a = \overline{P}(0) - \overline{P}(1) \geq 0$ and $b = \overline{P}(0) \geq 0$ are constants. This profile is obtained as a fixed point of system (1).

To motivative our problem, let us consider the vorticity

$$\omega(x,y,t) = U_y(x,y,t) - V_x(x,y,t)\ . \tag{5}$$

With (2) and (3), we get the equilibrium vorticity as

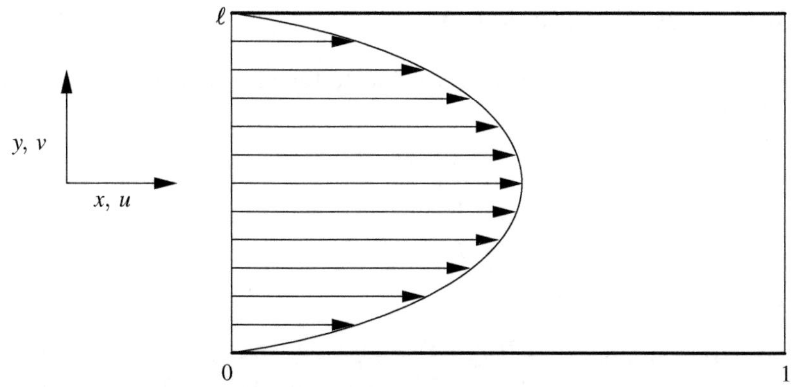

Fig. 1. 2D channel flow

$$\overline{\omega}\left(y\right) = \overline{U}'\left(y\right) - \overline{V}' = \frac{a}{2\nu}\left(\ell - 2y\right) . \qquad (6)$$

Suppose the vorticity at the walls is kept at its equilibrium values

$$\omega\left(x, 0, t\right) = \overline{\omega}\left(0\right) , \quad \omega\left(x, \ell, t\right) = \overline{\omega}\left(\ell\right) , \qquad (7)$$

and the wall–normal component of the velocity at the walls is zero:

$$V\left(x, 0, t\right) = 0 , \quad V\left(x, \ell, t\right) = 0 . \qquad (8)$$

The objective of these no–feedback boundary conditions might be the reduction of near–wall vorticity fluctuations. These boundary conditions imply

$$U_y\left(x, 0, t\right) = \omega\left(x, 0, t\right) + V_x\left(x, 0, t\right) = \frac{a\ell}{2\nu} , \qquad (9)$$

$$U_y\left(x, \ell, t\right) = \omega\left(x, \ell, t\right) + V_x\left(x, \ell, t\right) = -\frac{a\ell}{2\nu} . \qquad (10)$$

Under the boundary conditions (8)–(10), the Stokes equations

$$-\nu \Delta \mathbf{W} + \left(\mathbf{W} \cdot \nabla\right) \mathbf{W} + \nabla P = 0 , \qquad (11)$$

$$\operatorname{div} \mathbf{W} = 0 \qquad (12)$$

has a solution

$$U = \overline{U}\left(y\right) + C , \qquad (13)$$

$$V = \overline{V} , \qquad (14)$$

$$P = \overline{P}\left(x\right) , \qquad (15)$$

with an arbitrary constant C. This shows that under the boundary control (8)–(10) our objective of regulation to the equilibrium solution (2)–(3) can not be achieved. In more precise words, this solution is not asymptotically stable, and it can at best be marginally stable (with an eigenvalue at zero). To achieve *asymptotic* stabilization, in the next section we propose a feedback law which modifies the boundary condition (7).

3 Boundary Feedback Laws

In order to prepare for regulating the flow to the parabolic equilibrium profile (2)–(3), we set

$$u = U - \overline{U}\,, \tag{16}$$
$$v = V\,, \tag{17}$$
$$p = P - \overline{P}\,. \tag{18}$$

Then equation (1) becomes

$$\begin{cases} u_t - \nu\Delta u + uu_x + vu_y + \overline{U}u_x + \overline{U}'v + p_x = 0\,,\ 0 < x < 1,\ 0 < y < \ell,\ t > 0\,,\\ v_t - \nu\Delta v + uv_x + vv_y + \overline{U}v_x + p_y = 0\,,\ 0 < x < 1,\ 0 < y < \ell,\ t > 0\,,\\ u_x + v_y = 0\,,\ 0 < x < 1,\ 0 < y < \ell,\ t > 0\,,\\ u(x,y,0) = u_0,\ v(x,y,0) = v_0\,,\ 0 < x < 1,\ 0 < y < \ell, \end{cases} \tag{19}$$

To avoid dealing with an infinitely long channel, we assume that u, v, v_x and p are *periodic in the x–direction*, i.e.,

$$u(0,y,t) = u(1,y,t)\,,\ v(0,y,t) = v(1,y,t)\,,\quad 0 < y < \ell,\ t > 0\,, \tag{20}$$
$$v_x(0,y,t) = v_x(1,y,t)\,,\ p(0,y,t) = p(1,y,t)\,,\quad 0 < y < \ell,\ t > 0\,. \tag{21}$$

Our boundary control is applied via boundary conditions

$$\begin{cases} u(x,0,t) = ku_y(x,0,t)\,, & 0 < x < 1,\ t > 0\,,\\ u(x,\ell,t) = -ku_y(x,\ell,t)\,, & 0 < x < 1,\ t > 0\,,\\ v(x,0,t) = 0\,, & 0 < x < 1,\ t > 0\,,\\ v(x,\ell,t) = 0\,, & 0 < x < 1,\ t > 0\,, \end{cases} \tag{22}$$

where k is a positive constant. The physical implementation of this boundary condition is

$$U(x,0,t) = k\left[U_y(x,0,t) - \frac{a\ell}{2\nu}\right]\,, \tag{23}$$
$$U(x,\ell,t) = -k\left[U_y(x,\ell,t) + \frac{a\ell}{2\nu}\right]\,, \tag{24}$$
$$V(x,0,t) = 0\,, \tag{25}$$
$$V(x,\ell,t) = 0\,. \tag{26}$$

This means that we are actuating the flow velocity at the wall *tangentially*. Only the sensing of the wall shear stress $U_y(x,0,t)$ and $U_y(x,\ell,t)$ (at the respective points of actuation) is needed. The action of this feedback is pictorially represented in Fig. 2. The condition (23) and (24) can be also written as

$$U(x,0,t) = k\left[\omega(x,0,t) - \overline{\omega}(0)\right]\,, \tag{27}$$
$$U(x,\ell,t) = -k\left[\omega(x,\ell,t) - \overline{\omega}(\ell)\right]\,. \tag{28}$$

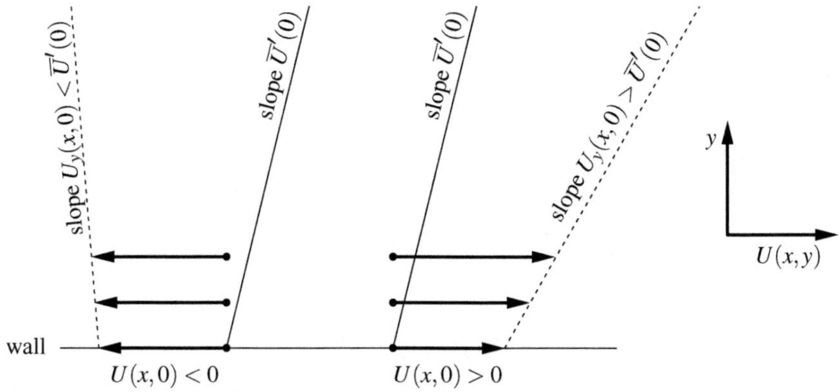

Fig. 2. Tangential velocity actuation

In the next sections we shall see that this control law achieves global asymptotic stabilization, whereas, as we saw in Sect. 2, the control law (7) is not asymptotically stabilizing.

4 The Result

Let $\Omega = (0,1) \times (0,\ell)$. In what follows, $H^s(\Omega)$ denotes the usual Sobolev space (see [24]) for any $s \in \mathbb{R}$. We shall be concerned with 2–dimensional vector function spaces and use boldface letters to denote them. For example we will use notations

$$\widetilde{\mathbf{V}} = \{(u,v) \in \widetilde{\mathbf{H}}^1 \,:\, u_x + v_y = 0 \text{ in } \Omega,\; v(x,0) = v(x,\ell) = 0\}, \tag{29}$$

$$\widetilde{\mathbf{H}} = \text{ the closure of } \widetilde{\mathbf{V}} \text{ in } \widetilde{\mathbf{L}}^2 \,. \tag{30}$$

From now on tilde sign refers to the periodicity in the x–direction. The norm in the space $\widetilde{\mathbf{V}}$ is defined by

$$\|\mathbf{w}\|_{\widetilde{\mathbf{V}}} = ((\mathbf{w},\mathbf{w}))^{1/2} \,, \tag{31}$$

where $((\cdot\,,\,\cdot))$ denotes the inner product of $\widetilde{\mathbf{V}}$ defined by

$$((\mathbf{w},\boldsymbol{\Phi})) = \int_0^\ell \int_0^1 Tr\left\{\nabla \mathbf{w}^T \nabla \boldsymbol{\Phi}\right\} dx dy + \frac{1}{k}\int_0^1 (u(x,0)\,\xi(x,0) + u(x,\ell)\,\xi(x,\ell))\, dx \,, \tag{32}$$

for all $\mathbf{w} = (u,v)$, $\boldsymbol{\Phi} = (\xi,\eta) \in \widetilde{\mathbf{V}}$.

Definition 1. *A function* $\mathbf{w} = (u,v) \in L^2([0,T];\widetilde{\mathbf{V}})$ *is a* weak solution *of system (19)–(22) if*

$$\frac{d}{dt}(\mathbf{w}, \boldsymbol{\Phi}) + \nu((\mathbf{w}, \boldsymbol{\Phi})) + ((\mathbf{w}\cdot\nabla)\mathbf{w}, \boldsymbol{\Phi}) + (\overline{U}\mathbf{w}_x, \boldsymbol{\Phi}) + \left(\overline{U}'v, \xi\right) = 0 \quad (33)$$

is satisfied for all $\boldsymbol{\Phi} = (\xi, \eta) \in \widetilde{\mathbf{V}}$ *and* $\mathbf{w}(x, y, 0) = \mathbf{w}_0(x, y)$ *for all* $(x, y) \in \Omega$.

Theorem 1. *Suppose that*

$$\nu > \sqrt{\frac{a\ell^3}{4}} \quad \text{and} \quad 0 < k < \ell/2\,, \quad (34)$$

and denote

$$\sigma = \frac{\nu}{\ell^2} - \frac{a\ell}{4\nu} > 0\,. \quad (35)$$

Then there exists a positive constant $c > 0$ *independent of* $\mathbf{w}_0$ *such that for the system (19) with periodic conditions (20)–(21), boundary control (22) and arbitrary initial data* $\mathbf{w}_0(x) \in \widetilde{\mathbf{H}}$, *there exists a unique weak solution* $\mathbf{w} \in L^2([0,\infty); \widetilde{\mathbf{V}}) \cap C([0,\infty); \widetilde{\mathbf{L}}^2)$ *that satisfies the global–exponential stability estimate*

$$\|\mathbf{w}(t)\| \le \|\mathbf{w}_0\| e^{-\sigma t} \quad \text{for all} \quad t \ge 0\,. \quad (36)$$

Solutions depend continuously on the initial data in the L^2*–norm and the existence, uniqueness and regularity statements hold for any* $\nu > 0$ *and* $k > 0$ *over finite time intervals.*

Remark 1. Similar statements hold for solutions in higher order ($L^2([0,\infty); \widetilde{\mathbf{H}}^2 \cap \widetilde{\mathbf{V}}) \cap L^\infty([0,\infty); \widetilde{\mathbf{V}})$ and $C^1([0,\infty); \widetilde{\mathbf{L}}^2) \cap C([0,\infty); \widetilde{\mathbf{H}}^2 \cap \widetilde{\mathbf{V}})$) spaces. In particular, it is possible to prove that

(1) The control inputs $u(x, 0, t)$ and $u(x, \ell, t)$ are bounded and go to zero as $t \to \infty$.
(2) The solution $\mathbf{w}(x, y, t)$ is continuous in all three arguments. This observation has an important practical consequence: the tangential velocity actuation at nearby points on the wall will be in the same direction.

Remark 2. If the viscosity $\nu \le \sqrt{\frac{a\ell^3}{2}}$, the problem of boundary control remains open. The methods presented in this paper can not be applied to this case and a radically different method needs to be developed.

5 Proof of Theorem

The proof of Theorem 1 and the statements mentioned in Remark 1 goes along the lines of the classical theory of Navier–Stokes equations [23, 28] with some modifications to accommodate the nonhomogeneous boundary conditions. Lyapunov techniques and Galerkin's method are used and several technical lemmas are employed. While we refer to [1] for details, here we prove the L^2 stability estimate (36). This is the first step in the whole proof and it also demonstrates the use of the energy method.

Let $\mathbf{w} = (u, v)$. We define the energy $E(\mathbf{w})$ of (19)–(22) as

$$E(\mathbf{w}) = \|\mathbf{w}\|^2 = \int_0^\ell \int_0^1 \left(u^2 + v^2\right) dxdy \,. \tag{37}$$

Multiplying the first equation of (19) by u and the second equation of (19) by v and integrating over Ω by parts, we obtain

$$\begin{aligned}
\dot{E}(\mathbf{w}) = & -2\nu \int_0^\ell \int_0^1 \left(u_x^2 + u_y^2 + v_x^2 + v_y^2\right) dxdy - 2 \int_0^\ell \int_0^1 \overline{U}' uv \, dxdy \\
& - \int_0^\ell u^3\Big|_{x=0}^1 dy - \int_0^1 u^2 v\Big|_{y=0}^\ell dx - \int_0^\ell \overline{U} u^2\Big|_{x=0}^1 dy \\
& - \int_0^\ell uv^2\Big|_{x=0}^1 dy - \int_0^1 v^3\Big|_{y=0}^\ell dx - \int_0^\ell \overline{U} v^2\Big|_{x=0}^1 dy \\
& -2 \int_0^\ell pu\Big|_{x=0}^1 dy - \int_0^1 pv\Big|_{y=0}^\ell dx + 2\nu \int_0^\ell u_x u\Big|_{x=0}^1 dy \\
& + 2\nu \int_0^1 u_y u\Big|_{y=0}^\ell dx + 2\nu \int_0^\ell v_x v\Big|_{x=0}^1 dy + 2 \int_0^1 v_y v\Big|_{y=0}^\ell dx \\
= & -2\nu \int_0^\ell \int_0^1 \left(u_x^2 + u_y^2 + v_x^2 + v_y^2\right) dxdy - 2 \int_0^\ell \int_0^1 \overline{U}' uv \, dxdy \\
& + 2\nu \int_0^1 u_y u\Big|_{y=0}^\ell dx \,.
\end{aligned} \tag{38}$$

Here we have used the relations

$$\begin{aligned}
& u_x(0, y, t) = u_x(1, y, t) \,, \quad u_y(0, y, t) = u_y(1, y, t) \,, \\
\text{and} \quad & v_y(0, y, t) = v_y(1, y, t) \,,
\end{aligned} \tag{39}$$

which follow from the periodic conditions (20)–(21) and the divergence free condition. Applying inequality

$$\int_0^\ell \int_0^1 \left(u^2 + v^2\right) dxdy \leq 2\ell \int_0^1 u^2(x, 0) \, dx + \ell^2 \int_0^\ell \int_0^1 \left(u_y^2 + v_y^2\right) dxdy \,, \tag{40}$$

which can be found in [1], it follows that

$$\begin{aligned}\dot{E}(\mathbf{w}) &\leq -\frac{2\nu}{\ell^2}E(\mathbf{w}) + \frac{4\nu}{\ell}\int_0^1 u^2(x,0,t)\,dx + \frac{\ell a}{2\nu}E(\mathbf{w}) \\ &\quad -\frac{2\nu}{k}\int_0^1 \left(u^2(x,\ell,t) + u^2(x,0,t)\right)dx \\ &= -\frac{2\nu}{\ell^2}E(\mathbf{w}) + \frac{\ell a}{2\nu}E(\mathbf{w}) \\ &\quad -\int_0^1 \left(2\nu\left(\frac{1}{k} - \frac{2}{\ell}\right)u^2(x,0,t) + \frac{2\nu}{k}u^2(x,\ell,t)\right)dx \\ &\leq -\left(\frac{2\nu}{\ell^2} - \frac{a\ell}{2\nu}\right)E(\mathbf{w}) . \end{aligned} \tag{41}$$

This implies the global–exponential stability in L^2-norm, i.e. (36).

Remark 3. The boundary integral

$$\int_0^1 \left(2\nu\left(\frac{2}{\ell} - \frac{1}{k}\right)u^2(x,0,t) - 2\nu\frac{1}{k}u^2(x,\ell,t)\right)dx \tag{42}$$

in (41) is negative even for large Reynolds numbers (small kinematic viscosity) if k is sufficiently small. Hence, it improves the stability properties in general. The trace theorem however does not allow us to compare this term and the total energy and to prove the stability results for large Reynolds numbers. This shows the need for numerical simulation.

6 Numerical Simulation

The simulation example in this section is performed in a channel of length 4π and height 2 for Reynolds number $Re = 15000$ ($a = 2/15000$, $\nu = 1/15000$), which is five orders of magnitude greater than required in Theorem 1, and is three times the critical value (5772, corresponding to loss of linear stability) for 2D channel flow. The validity of the stabilization result beyond the assumptions of Theorem 1 is not completely surprising since our Lyapunov analysis is based on conservative energy estimates.The control gain used is $k = 1$.

A hybrid Fourier pseudospectral–finite difference discretization and the fractional step technique based on a hybrid Runge–Kutta/Crank–Nicolson time discretization was used to generate the results [3, 1, 5]. The number of grid points used in our computations was 128×120 and the (adaptive) time step was in the range of $0.05 - 0.07$. The grid points had hyperbolic tangent ($y_j = 1 + \tanh\left(s\left(2\frac{j}{NY} - 1\right)\right)/\tanh(s)$ $j = 0, \ldots, NY$) distribution with stretching factor $s = 1.75$ in the vertical direction in order to achieve high resolution in the critical boundary layer. In order to obtain the flow at the walls in the controlled case the quadratic Three–Point Endpoint Formula was used to approximate the derivatives at the boundary ($U_y(x,0,t)$, $U_y(x,\ell,t)$).

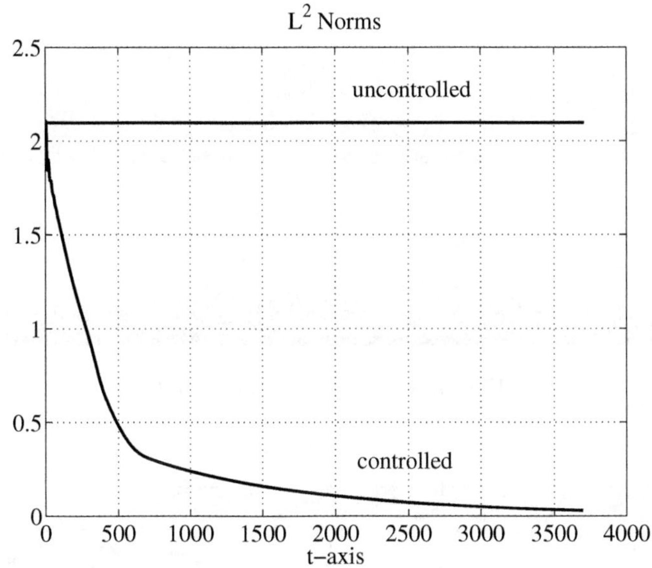

Fig. 3. Energy Comparison

This formula is applied in a semi-implicit way in order to avoid numerical instabilities. The numerical results show very good agreement with results obtained from a finite volume code used at early stages of simulations. As initial data we consider a statistically steady state flow field obtained from a random perturbation of the parabolic profile over a large time period using the uncontrolled system. Figure 3 shows that our controller achieves stabilization. This is expressed in terms of the L^2–norm of the error between the steady state and the actual velocity field, the so called perturbation energy, which corresponds to system (19)–(22) with $k = 0$ (zero Dirichlet boundary conditions on the walls) in the uncontrolled case. The initially fast perturbation energy decay somewhat slows down for larger time. What we see here is an interesting example of interaction between linear and nonlinear behavior in a dynamical system. Initially, when the velocity perturbations are large, and the flow is highly nonlinear (exhibiting Tollmien–Schlichting waves with recirculation, see the uncontrolled flow in Figs. 4 and 5). The strong convective (quadratic) nonlinearity dominates over the linear dynamics and the energy decay is fast. Later, at about $t = 500$, the recirculation disappears, the controlled flow becomes close to laminar, and linear behavior dominates, along with its exponential energy decay (with small decay rate).

In the vorticity map, depicted in Fig. 4 it is striking how uniform the vorticity field becomes for the controlled case, while we observe quasi–periodic bursting (cf. [19]) in the uncontrolled case. We obtained similar vorticity maps of the uncontrolled flow for other (lower) Reynolds numbers, that show agreement qualitatively with the vorticity maps obtained by Jiménez [19]. His paper

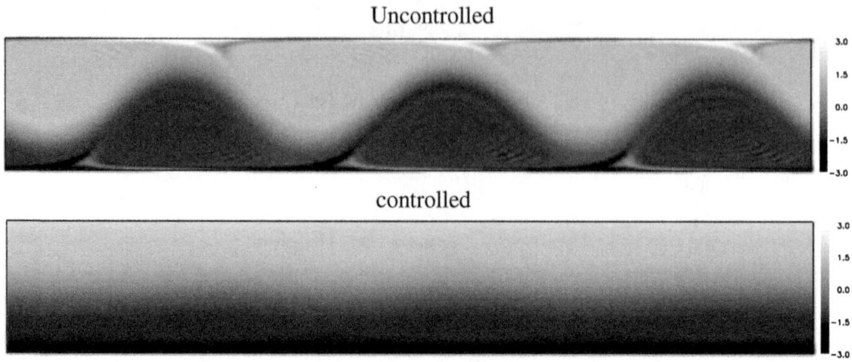

Fig. 4. Vorticity Maps at $t = 700$

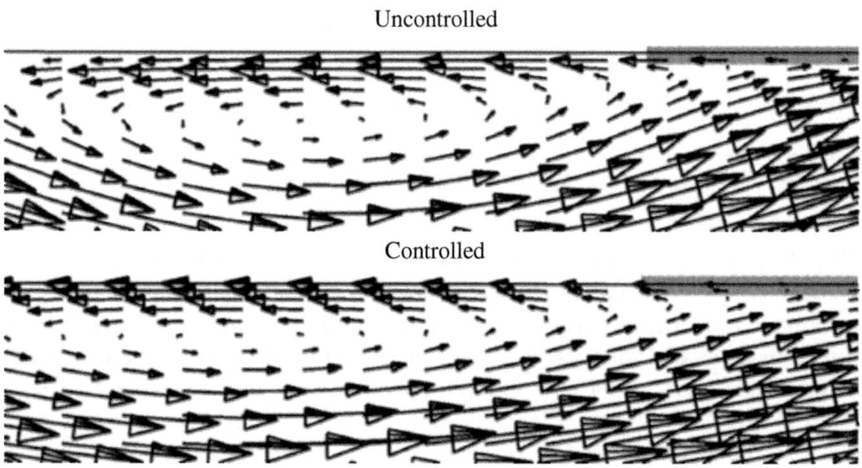

Fig. 5. Recirculation in the flow at $t = 120$, in a rectangle of dimension 1.37×0.31 zoomed out of a channel of dimension $4\pi \times 2$. The shaded region (upper right corner) is magnified in Fig. 6

explains the generation of vortex blobs at the wall along with their ejection into the channel and their final dissipation by viscosity in the uncontrolled case.

The uniformity of the wall shear stress ($U_y|_{\text{wall}}$) in the controlled flow can be also observed in Fig. 6. Our boundary feedback control (tangential actuation) adjusts the flow field near the upper boundary such that the controlled wall shear stress almost matches that of the steady state profile. The region is at the edge of a small recirculation bubble (Fig. 5) of the uncontrolled flow, hence there are some flow vectors pointing in the upstream direction while others are oriented downstream. The time is relatively short ($t = 120$) after the introduction of the control and the region is small. As a result it is still possible to see actuation both downstream and upstream. Nevertheless the

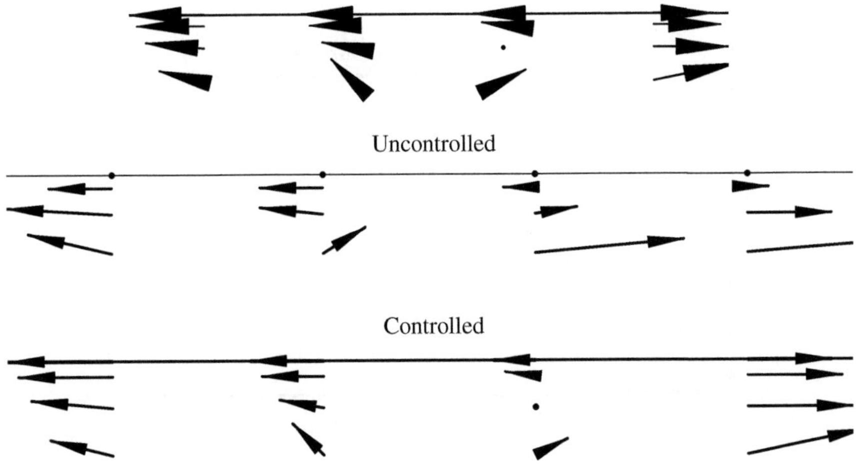

Fig. 6. Velocity field in a rectangle of dimension 0.393×0.012 zoomed out of a channel of dimension $4\pi \times 2$, at time $t = 120$. The control (thick arrows) acts both *downstream* and *upstream*. The control maintains the value of shear near the desired (laminar) steady–state value

controlled velocity varies continuously. Figure 5 shows that the effect of control is to smear the vortical structures out in the streamwise direction. It is well known that in wall bounded turbulence instabilities are generated at the wall. In two dimensional flows these instabilities are also confined to the walls. As a result, our control effectively stabilizes the flow.

We obtain approximately 71% drag reduction (see Fig. 7) as a byproduct of our special control law. The drag in the controlled case "undershoots" bellow the level corresponding to the laminar flow and eventually agrees with it up to two decimal places. It is striking that even though drag reduction was not an explicit control objective (as in most of the works in this field), the stabilization objective results in a controller that reacts to the wall shear stress error, and leads to an almost instantaneous reduction of drag to the laminar level.

7 Simulations with Only Parts of the Wall Controlled

While the use of boundary control is practically more feasible than distributed control, it is even more realistic to use control applied only on part of the boundary wall. In the present section we compare several different configurations of the channel flow with boundary control restricted to different parts of the wall (see Fig. 8). The subintervals (patches) are of equal length and their number on each wall is one, two, four or eight. In addition we consider the

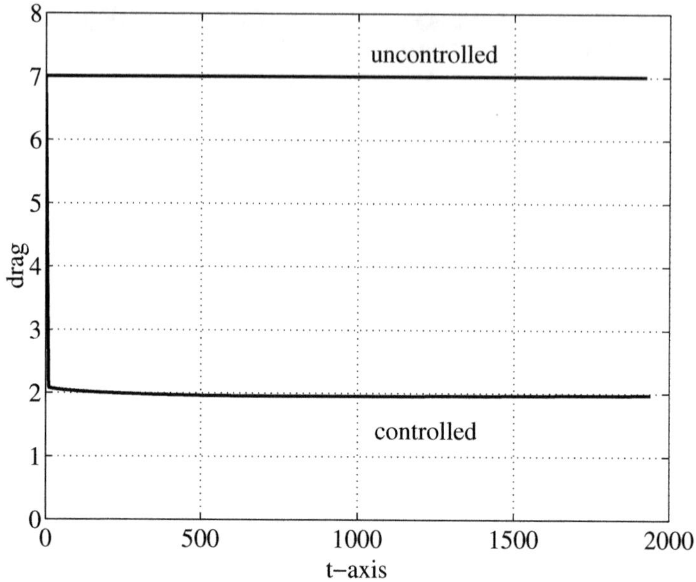

Fig. 7. Instantaneous drag

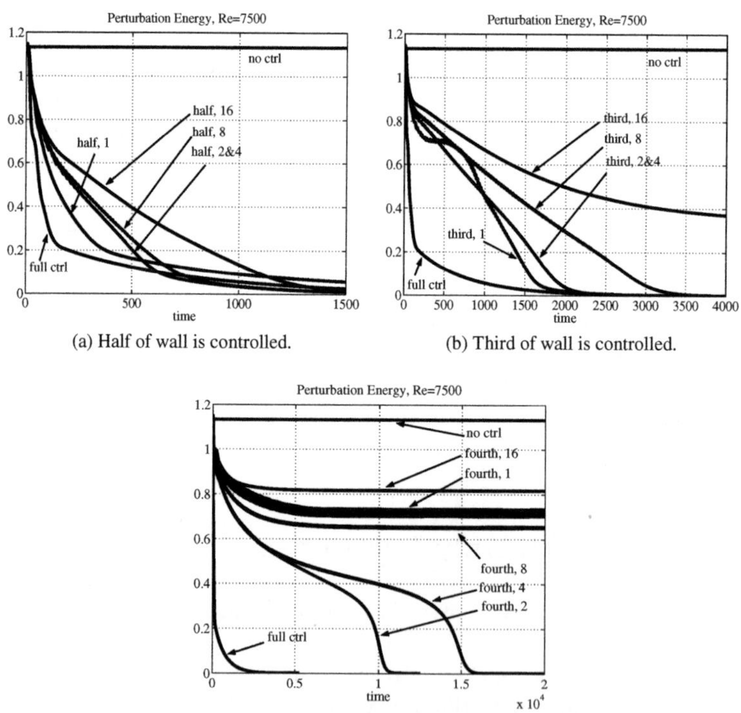

(a) Half of wall is controlled.

(b) Third of wall is controlled.

(c) Fourth of wall is controlled.

Fig. 8. Configuration of control patches

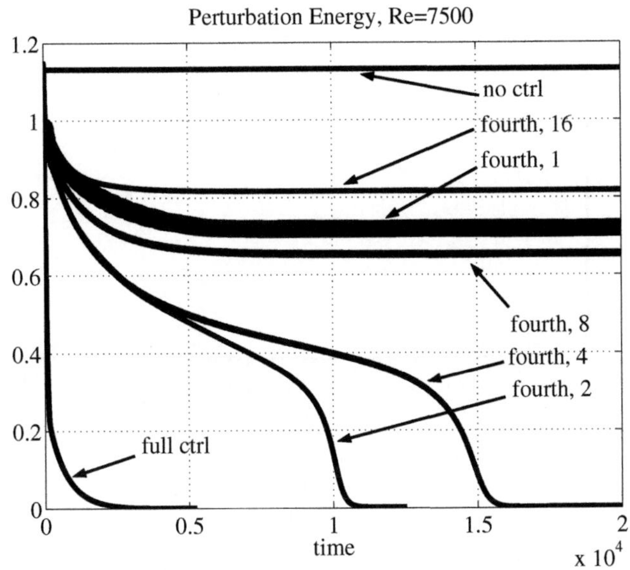

Fig. 9. Comparison of perturbation energies (I)

case when control is applied only on one wall. The total length of the patches is one fourth, one third or one half of the total wall length. In all cases the control gain k introduced in (23) and (24) has been set to the value one. For these cases we present perturbation energy comparison, including the fully controlled and the uncontrolled case. Our general observations based on Figs. 9 and 10 are the following.

The flow can be laminarized even when the size of the controlled region is small relative to the uncontrolled region. The smallest patch size we achieved stability with was the 1/4 case (Fig. 9, part (c)).

We achieve better performance with control concentrated in one area of the two walls than with many small patches distributed evenly along the walls (Fig. 9). With 16 patches stability was not achieved in the 1/3 case (Fig. 9, part (b)). The corresponding curve levels around the value 0.33 right after $t = 4000$. In the 1/4 case (Fig. 9, part (c)) already 8 patches degraded the control to a level where laminarization was not possible.

In the 1/2 and 1/3 cases concentrating all the control on one wall does not result in significantly different control performance than the one–paired configuration (Fig. 9, parts (a)–(b)). In the 1/4 case control restricted to one wall does not fully stabilize the parabolic profile, while one and two pairs of control patches do stabilize.

In a given patch configuration larger control surface achieves stability faster. The controlled and uncontrolled parts of the wall are competing against

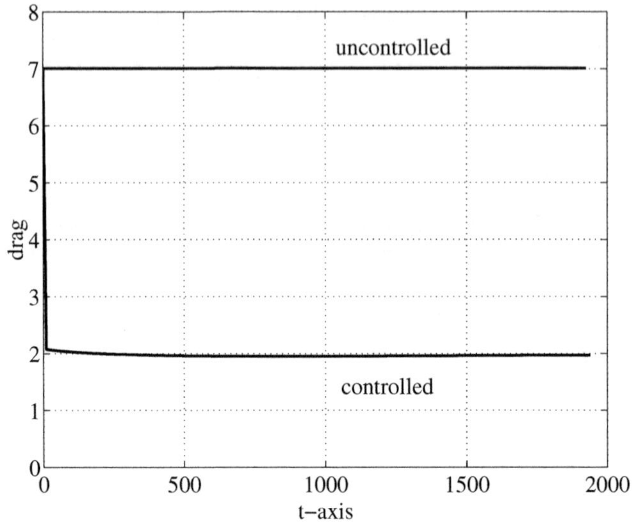

Fig. 10. Comparisons of perturbation energies (II)

each other. The controlled part has to be large enough so that its effect has time to destroy the wall bounded turbulent structures, and the uncontrolled part has to be small enough so that instabilities do not grow over that region more than they decay over the controlled region (Fig. 10).

Acknowledgment

We thank Thomas Bewley and Peter Blossey for their generous help with the numerical part of this work and continuous exchange of ideas, and we thank Javier Jiménez for his helpful comments.

References

[1] A. Balogh, W.-J. Liu and M. Krstic, "Stability Enhancement by Boundary Control in 2D Channel Flow," *IEEE Transactions on Automatic Control*, vol. 46, no. 11, pp. 1696–1711, 2001.
[2] V. Barbu, "The time optimal control of Navier–Stokes equations," *Systems Control Lett.*, vol. 30, no. 2–3, pp. 93–100, 1997.
[3] T.R. Bewley, P. Moin and R. Temam, "DNS–based predictive control of turbulence: an optimal benchmark for feedback algorithms," *J. Fluid Mech.*, vol. 447, pp. 179–225, 2001.
[4] T.R. Bewley, R. Temam and M. Ziane, "A general framework for robust control in fluid mechanics," *Physica D, vol.* 138, pp. 360–392, 2000.
[5] M.H. Carpenter and C.A. Kennedy, "Fourth-order 2N Runge-Kutta schemes," *NASA technical memorandum,* no. 109112, 1994.

[6] J.-M. Coron, "On the controllability of the 2–D incompressible Navier–Stokes equations with the Navier slip boundary conditions," *ESAIM: Control, Optim. Cal. Var.*, vol. 1, pp. 35–75, 1996.

[7] J.-M.Coron, On the null asymptotic stabilization of the two–dimensional incompressible Euler equations in a simply connected domain," *SIAM Journal on Control and Optimization*, vol. 37, No. 6, pp. 1874–1896, 1999.

[8] L. Cortelezzi, K.H. Lee, J. Kim and J.L. Speyer, "Skin–Friction Drag Reduction Via Robust Reduced–Order Linear Feedback Control", *International Journal of Computational Fluid Dynamics*, Vol. 11, No. 1-2, 79-92, 1998.

[9] M. Desai and K. Ito, "Optimal controls of Navier–Stokes equations," *SIAM J. Control Optim.*, vol. 32, no. 5, pp. 1428–1446, 1994.

[10] H.O. Fattorini and S.S. Sritharan, "Optimal control problems with state constraints in fluid mechanics and combustion," *Appl. Math. Optim.*, vol. 38, no. 2, pp. 159–192, 1998.

[11] E. Fernández–Cara, "On the approximate and null controllability of the Navier–Stokes equations," *SIAM Rev.*, vol. 41, no. 2, pp. 269–277, 1999.

[12] A.V. Fursikov, "Exact boundary zero controllability of three–dimensional Navier–Stokes equations," *J. Dynam. Control Systems*, vol. 1, no. 3, pp. 325–350, 1995.

[13] A.V. Fursikov, M.D. Gunzburger and L.S. Hou, "Boundary value problems and optimal boundary control for the Navier–Stokes system: the two–dimensional case." *SIAM J. Control Optim.*, vol. 36, no. 3, pp. 852–894, 1998.

[14] A.V. Fursikov and O.Y. Imanuvilov, "On exact boundary zero–controllability of two–dimensional Navier–Stokes equations, Mathematical problems for Navier–Stokes equations (Centro, 1993)," *Acta Appl. Math.*, vol. 37, no. 1–2, pp. 67–76, 1994.

[15] M.D. Gunzburger, L.S. Hou, T.P. Svobodny, "Boundary velocity control of incompressible flow with an application to viscous drag reduction," *SIAM J. Control Optim.*, vol. 30, no. 1, pp. 167–181, 1992.

[16] L.S Hou and Y. Yan, "Dynamics and approximations of a velocity tracking problem for the Navier–Stokes flows with piecewise distributed control," *SIAM J. Control Optim.*, vol. 35, no. 6, pp. 1847–1885, 1997 .

[17] O.Y. Imanuvilov, "On exact controllability for the Navier–Stokes equations," *ESAIM: Control, Optim. Cal. Var.*, vol. 3, pp. 97–131, 1998.

[18] K. Ito and S. Kang, "A dissipative feedback control synthesis for systems arising in fluid dynamics," *SIAM J. Control Optim.*, vol. 32, no. 3, pp. 831–854, 1994.

[19] J. Jiménez, "Transition to turbulence in two–dimensional Poiseuille flow," *J. Fluid Mech.*, vol. 218, pp. 265–297, 1990.

[20] S.S. Joshi, J.L. Speyer and J. Kim, "A system theory approach to the feedback stabilization of infinitesimal and finite–amplitude disturbances in plane Poiseuille flow," *Journal of Fluid Mechanics*, vol. 332, pp. 157–184, 1997.

[21] S.S. Joshi, J.L. Speyer and J. Kim, "Finite dimensional optimal control of Poiseuille flow," *Journal of Guidance, Control, and Dynamics*, vol. 22, no. 2, pp. 340–348, 1999.

[22] L.R. Keefe, "Method and apparatus for reducing the drag of flows over surfaces," US Patent US5803409, 1998.

[23] O.A. Ladyzhenskaya, *The Mathematical Theory of Viscous Incompressible Flow*, Second English edition, Gordon and Breach, Science Publisher, Inc., New York, 1969.

[24] J.L. Lions and E. Magenes, *Non–homogeneous Boundary value Problems and Applications, Vol.I,* Springer–Verlag, Berlin, Heidelberg, New York, 1972.

[25] B.L. Rozhdestvensky and I.N. Simakin, "Secondary flows in a plane channel: their relationship and comparison with turbulent flows," *J. Fluid Mech.*, vol. 147, pp. 261–289, 1984.

[26] B.L. Smith and A. Glezer, "The formation and evolution of synthetic jets," *Phys. Fluids,* vol. 10, no. 9, pp. 2281–2297, 1998.

[27] S.S. Sritharan, "Dynamic programming of the Navier–Stokes equations," *Systems Control Lett.,* vol. 16, no. 4, pp. 299–307, 1991.

[28] R. Temam, *Navier–Stokes Equations: Theory and Numerical Analysis,* Third (Revised) edition, North–Holland Publishing Company, Amsterdam, 1984.

[29] R. Temam, *Navier–Stokes Equations and Nonlinear Functional Analysis,* Second edition, SIAM, Philadelphia, 1995.

Exact Controllability and Feedback Stabilization from a Boundary for the Navier–Stokes Equations

Andrei V. Fursikov

Department of Mathematics and Mechanics, Moscow State University, 119899 Moscow, Russia

Abstract. For $2D$ Navier–Stokes equations defined in a bounded domain Ω we study stabilization of solution near a given steady-state flow $\hat{v}(x)$ by means of feedback control defined on a part Γ of boundary $\partial\Omega$. New mathematical formalization of feedback notion is proposed. With its help for a prescribed number $\sigma > 0$ and for an initial condition $v_0(x)$ placed in a small neighbourhood of $\hat{v}(x)$ a control $u(t, x')$, $x \in \Gamma$, is constructed such that solution $v(t, x)$ of obtained boundary value problem for 2D Navier–Stokes equations satisfies the inequality:

$$\|v(t, \cdot) - \hat{v}\|_{H^1} \leqslant ce^{-\sigma t} \text{ for } t \geqslant 0 \,.$$

1 Introduction

In this paper we study stablization problem for two-dimensional (2D) Navier–Stokes equations defined in a bounded domain $\Omega \subset \mathbb{R}^2$ which is controlled by Dirichlet boundary condition for velocity vector field. Let $(\hat{v}(x), \nabla\hat{p}(x))$, $x \in \Omega$ be a steady-state solution for Navier–Stokes equations. Let suppose that $\hat{v}$ is an unstable singular point for the dynamic system generated by evolutionary Navier-Stokes system supplied with zero condition $v|_{\partial\Omega} = 0$ on the boundary $\partial\Omega$ of Ω.

The stabilization problem is as follows: Given $\sigma > 0$ and initial condition $v_0(x)$ for evolutionary Navier–Stokes system placed in a small neighbourhood of $\hat{v}$ in H^1, find Dirichlet boundary condition (control) $u(t, x')$ defined on $\mathbb{R}_+ \times \partial\Omega$ such that the solution $v(t, x)$ of obtained boundary value problem satisfies inequality

$$\|v(t, \cdot) - \hat{v}(\cdot)\|_{H^1(\Omega)} \leqslant ce^{-\sigma t} \text{ as } t \to \infty \,. \tag{1}$$

Actually, the stabilization result in such formulation follows immediately from the exact controllability result of [FE],[F1] which is formulated in Sect. 1. But below we impose on desired control very important additional property: u has to be a feedback control.

Problem of stabilization by boundary feedback control was studied earlier mostly for hyperbolic equations and related to them systems (see, for instance, [Li], [Lag], [C]). Some results, connected with Burgers equation was obtained also ([BK]). (Of course, we do not pretend on completeness of references.)

In [F2] new mathematical formalization of feedback property was proposed and stabilization problem for quasilinear parabolic equation by feedback boundary control was solved. Here we obtain solution for problem of stabilization by boundary feedback control for two-dimensional Navier–Stokes system when a control is concentrated on a part of boundary. Complete proof of this result will be exposed in [F3]

2 Exact Controllability from the Boundary of the Navier–Stokes System

Let $\Omega \subset R^d, d = 2, 3$, be a bounded domain with a boundary $\partial\Omega$ of the class C^∞, $T > 0$, $Q_T = (0,T) \times \Omega, \Sigma_T = (0,T) \times \partial\Omega$. We consider in Q_T the mixed boundary value problem for the Navier–Stokes system:

$$\partial_t v(t,x) - \Delta v + (v,\nabla)v + \nabla p = f(t,x), \quad \operatorname{div} v(t,x) = 0\ , \tag{2}$$

$$v(t,x)\,|_{t=0} = v_0(x)\ , \tag{3}$$

$$v\,|_{\Sigma_T} = u\ , \tag{4}$$

where $\partial_t = \frac{\partial}{\partial t}$, $x = (x_1,\dots,x_d) \in \Omega$, $v(t,x) = (v_1(t,x),\dots,v_d(t,x))$ is the velocity vector field of a fluid flow, $\nabla p(t,x)$ is the pressure gradient of a liquid, $f(t,x) = (f_1(t,x),\dots f_d(t,x))$ is a given density of exterior forces, Δ is the Laplace operator, $(v,\nabla)v = \sum_i v_i\partial_i v$, $\partial_i = \frac{\partial}{\partial x_i}$, v_0 is an initial vector field. The vector field u defined on the lateral surface Σ_T is not given: it is a control.

The exact controllability problem from the boundary for the Navier–Stokes system is as follows: given a solution $(\hat{v}(t,x), \hat{p}(t,x))$ of (2):

$$\partial_t\hat{v}(t,x) - \Delta\hat{v} + (\hat{v},\nabla)\hat{v} + \nabla\hat{p} = f(t,x), \quad \operatorname{div}\hat{v} = 0\ , \tag{5}$$

find a control $u(t,x)$ defined on Σ_T such that the solution $v(t,x)$ of problem (2)–(4) coincides at instant $t = T$ with $\hat{v}(T,x)$:

$$v(t,x)|_{t=T} \equiv \hat{v}(T,x)\ . \tag{6}$$

We introduce the functions spaces. As usually $H^k(\Omega)$ where k is a natural number is the Sobolev space of scalar functions, defined and square integrable on Ω together with all its derivatives up to order k and $(H^k(\Omega))^d$ is the analogous Sobolev space of vector fields. We set

$$V^k(\Omega) = \{v(x) = (v_1,\dots v_d) \in (H^k(\Omega))^d : \quad \operatorname{div} v = 0\}\ , \tag{7}$$

$$H^{1,2(k)}(Q) = \left\{v \in L_2(0,T; H^{2+k}(\Omega)) : \partial_t v \in L_2(0,T; H^k(\Omega))\right\}\,, \tag{8}$$

$$V^{1,2(k)}(Q) = \left\{v \in (H^{1,2(k)}(Q))^d : \operatorname{div} v = 0\right\}\,, \tag{9}$$

$$H^{1,2}(Q) = H^{1,2(0)}(Q), \quad V^{1,2}(Q) = V^{1,2(0)}(Q)\,.$$

Result on exact controllability is as follows:

Theorem 2.1. *Let $f \in L_2(0,T; V^2(\Omega)), v_0 \in V^3(\Omega)$ and a solution $(\hat{v}, \hat{p}) \in V^{1,2(2)}(Q) \times L_2(0,T; H^3(\Omega))$ of (5) be given. We suppose that over each connected componenet $\partial\Omega_i$ of the boundary $\partial\Omega = \cup_{i=1}^M \partial\Omega_i$ the equalities hold:*

$$\int\limits_{\partial\Omega_i} \hat{v}(t,x') \cdot \nu(x') ds = 0, \quad a.a \quad t \in [0,t], \qquad \int\limits_{\partial\Omega_i} v_0(x') \cdot \nu(x') ds = 0 \tag{10}$$

where ν is the vector field of outward normals to $\partial\Omega$. Then there exists a solution $(v,p,u) \in V^{1,2}(Q) \times L_2(0,T; H^1(\Omega)) \times (L_2(0,T; H^{3/2}(\partial\Omega)))^d$ of problem (2)–(4), (6).

Stabilization problem can be regarded as a special case of exact controllability problem presented above. Indeed, suppose that the right side f from (2) does not depend on t and a given solution from formulation of exact controllability problem is steady-state, i.e. a pair $(\hat{v}(x), \hat{p}(x))$ satisfies the equations:

$$-\Delta\hat{v} + (\hat{v}, \nabla)\hat{v} + \nabla\hat{p} = f(x), \quad \operatorname{div}\hat{v} = 0\,. \tag{11}$$

Then exact controllability problem is called stabilization problem and we get the followig evident corollary of Theorem 2.1:

Theorem 2.2. *Let $f \in V^2(\Omega), v_0 \in V^3(\Omega)$ and a solution $(\hat{v}, \hat{p}) \in V^4(\Omega) \times H^3(\Omega)$ of (11) be given. Suppose that equalities (10) hold. Then there exists a solution*

$$(v,p,u) \in V^{1,2}(Q) \times L_2(0,T; H^1(\Omega)) \times (L_2(0,T; H^{3/2}(\partial\Omega)))^d$$

of problem (2)–(4), satisfying

$$v(T,x) = \hat{v}(x)\,. \tag{12}$$

We do not prove here Theorems 2.1, 2.2: their complete proof is given in [FE]. Nevertheless we want to note that solution of exact controllability problem is reduced to solution of ill-posed problem (2), (3), (6) with unknown functions $(v(t,x), \nabla p(t,x))$. After solution of this il-posed problem we get control u with help of (5). Solution of stabilization problem in formulation given above is reduced to ill-posed problem (2), (3), (12) by just the same way.

Thus, Theorems 2.1, 2.2 are existence theorems for ill-posed problems. The method for construction of control u in these theorems is such that in order to construct the control $u(t,x')$ even for $t \in (0,t_0)$ with $t_0 \ll T$ we use information on problem (2), (3), (6) and on its solution $v(t,x)$ on the whole time interval $(0,T)$. Therefore if at certain $t_0 \in (0,T)$ some unpredictable fluctuation of solution $v(t,x)$ would arise, control u will not react on this fluctuation and in virtue of ill-posedness of (2), (3), (6), solution of this problem will be quickly completely destroyed by this fluctuation. Unpredictable fluctuations definitely arise in numerical simulations because all numbers in computers are set approximately. This shows how important to be able to construct a control that can react on unpredictable fluctuations of solution suppressing them, i.e. to construct feedback control.

Below we study the problem of stabilization by boundary feedback control for 2D Navier–Stokes equations. Imposing feedback property on boundary control we simultaneously weaken condition (12) and change it on exponential decay (1.1) with prescribed $\sigma > 0$.

3 Ozeen Equations

We begin investigation of stabilization problem from the case of linearized Navier–Stokes equations, i.e. from the Ozeen equations.

3.1 Formulation of the Problem

Let $\Omega \subset \mathbb{R}^2$, $\partial\Omega \in C^\infty$, $Q = \mathbb{R}_+ \times \Omega$. We consider the Ozeen equations:

$$\partial_t v(t,x) - \Delta v + (a(x),\nabla)v + (v,\nabla)a + \nabla p(t,x) = 0\,, \tag{13}$$

$$\operatorname{div} v(t,x) = 0 \tag{14}$$

with initial condition

$$v(t,x)|_{t=0} = v_0(x)\,. \tag{15}$$

Here $(t,x) = (t,x_1,x_2) \in Q$, $v(t,x) = (v_1,v_2)$, $a(x) = (a_1(x),a_2(x))$ is a given solenoidal vector field ($\operatorname{div} a = 0$).

We suppose that the boundary $\partial\Omega$ of Ω is decomposed on two parts:

$$\partial\Omega = \overline{\Gamma} \cup \overline{\Gamma}_0, \quad \Gamma \neq \emptyset \tag{16}$$

where Γ, Γ_0 are open sets (in topology of $\partial\Omega$). Here, as usual, the line above means the closer of a set. We define $\Sigma = \mathbb{R}_+ \times \Gamma$, $\Sigma_0 = \mathbb{R}_+ \times \Gamma_0$, and we set:

$$v|_{\Sigma_0} = 0, \quad v|_\Sigma = u \tag{17}$$

where u is a control, concentrated on Σ.

Let a magnitude $\sigma > 0$ be given. The problem of stabilization with the rate σ of a solution to problem (13)–(15), (17) is to construct a control u on Σ such that the solution $v(t,x)$ of boundary value problem (13)–(15), (17) satisfies:

$$\|v(t,\,\cdot\,)\|^2_{L_2(\Omega)} \leqslant ce^{-\sigma t} \quad \text{for } t \geqslant 0 \tag{18}$$

where $c > 0$ depends on v_0, σ and Γ_0. Moreover, we require that this control u satisfies the feedback property in the meaning indicatad below.

Let us give the exact formulation of this feedback property. Let $\omega \subset \mathbb{R}^2$ be a bounded domain such that

$$\Omega \cap \omega = \emptyset, \quad \overline{\Omega} \cap \overline{\omega} = \overline{\Gamma}\,. \tag{19}$$

We set

$$G = \operatorname{Int}(\overline{\Omega} \cup \overline{\omega}) \tag{20}$$

(the denotion $\operatorname{Int} A$ means, as always, the interior of the set A). We suppose that $\partial G \in C^\alpha$ and in all points except $\overline{\Gamma} \setminus \Gamma \equiv \partial \Gamma$ it possesses the C^∞ smoothness. The exact conditions on α will be imposed below in an appropriate place.

We extend problem (13)–(15) from Ω up to G. Let us assume that

$$a(x) \in V^2(G) \cap \left(H^1_0(G)\right)^2\,. \tag{21}$$

That is why the extension of (13)–(15) from Ω to G can be written as follows:

$$\partial_t w(t,x) - \Delta w + (a(x), \nabla)w + (w, \nabla)a + \nabla p(t,x) = 0, \quad \operatorname{div} w = 0\,, \tag{22}$$

$$w(t,x)|_{t=0} = w_0(x), \qquad w|_S = 0 \tag{23}$$

where $S = \mathbb{R}_+ \times \partial\Omega$ (we impose on w the zero Dirichlet boundary condition). Note that actually w_0 from (23) will be some special extension of $v_0(x)$ from (15).

For vector fields defined on G we denote by γ_Ω the operator of restriction on Ω and by γ_Γ we denote the operator of restriction on Γ:

$$\gamma_\Omega : V^k(G) \longrightarrow V^k(\Omega), \quad \gamma_\Gamma : V^k(G) \longrightarrow V^{k-1/2}(\Gamma),\ k \geqslant 0\,. \tag{24}$$

Evidently, operators (24) are bounded.

Definition 3.1. *A control $u(t,x)$ in (13)–(15), (17) is called feedback if*

$$v(t,\,\cdot\,) = \gamma_\Omega w(t,\,\cdot\,), \quad u(t,\,\cdot\,) = \gamma_\Gamma w(t,\,\cdot\,) \qquad \forall t \geqslant 0 \tag{25}$$

where $(v(t,\,\cdot\,), u(t,\,\cdot\,))$ is the solution of stabilization problem (13)–(15), (17) and $w(t,\,\cdot\,)$ is the solution of boundary value problem (22), (23).

Below we prove that for given $\sigma > 0, v_0$ the problem (13)–(15), (17) can be stabilized with help of feedback control in the meaning of Definition 3.1.

3.2 Preliminaries

Let G be domain (20) and $V_0^0(G) = \{v(x) \in V^0(G) : v \cdot \nu|_{\partial\Omega} = 0\}$ where $V^0(G)$ is space (2.7) with $k = 0$, $\nu(x)$ is the vector-field of outer normals to ∂G. Denote by $\pi : (L_2(G))^2 \longrightarrow V_0^0(G)$ the operator of ortogonal projection. We consider the Ozeen steady state operator

$$Av \equiv -\pi\Delta v + \pi[(a(x), \nabla)v + (v, \nabla)a] : V_0^0(G) \longrightarrow V_0^0(G) \tag{26}$$

where $a(x)$ is vector-field (21). This operator is closed and has the domain $\mathcal{D}(A) = V^2(G) \cap (H_0^1(G))^2$ which is dense in $V_0^0(G)$.

Asumming that space in (26), is complex we denote by $\rho(A)$ the resolvent set of operator A, i.e. the set of $\lambda \in \mathbb{C}$ such that the resolvent operator $R(\lambda, A) \equiv (\lambda I - A)^{-1} : V_0^0(G) \longrightarrow V_0^0(G)$ is defined and continuous. Here I is identity operator. Denote by $\Sigma(A) \equiv \mathbb{C}^1 \setminus \rho(A)$ the spectrum of operator A.

As well-known, for $\lambda \in \rho(A)$ resolvent $R(\lambda, A)$ is a compact operator, and the spectrum $\Sigma(A)$ consists of a discret set of points.

We decompose the resolvent $R(\lambda, -A)$ in a neighbourhood of $-\lambda_j \in \Sigma(-A)$:

$$R(\lambda, -A) = \sum_{k=-m}^{\infty} (\lambda + \lambda_j)^k R_k, \quad R_k = \int\limits_{|\lambda+\lambda_j|=\varepsilon} \frac{R(\lambda, -A)}{2\pi i(\lambda + \lambda_j)^{k+1}} d\lambda \,. \tag{27}$$

Note that $m < \infty$.

Let A^* be the operator formally adjoint to Ozeen operator (26). Evidently, A^* is a closed operator with domain $\mathcal{D}(A^*) = V^2(G) \cap (H_0^1(G))^2$. Moreover, operator A^* is sectorial with a compact resolvent and

$$\rho(A^*) = \overline{\rho(A)}, \qquad R(\lambda, A)^* = R(\bar{\lambda}, A^*) \quad \forall\, \lambda \in \rho(A) \tag{28}$$

(Here the line above means complex conjugation.) Below we always assume that

$$\text{vector field } a(x) \text{ from } (26), (21) \text{ is real valued} \,. \tag{29}$$

That is why we have $\rho(A) = \overline{\rho}(A) = \rho(A^*) = \overline{\rho}(A^*)$.

Let $-\lambda_j \in \Sigma(-A)$ be an eingenvalue of $-A$, and $e \neq 0$, $e \in \ker(\lambda_0 I + A)$ be an eingenvector. Vector e_k is called joined vector of order k to e if e_k satisfies:

$$(\lambda_0 I + A)e = 0, \quad e + (\lambda_0 I + A)e_1 = 0, \ldots, \quad e_{k-1} + (\lambda_0 I + A)e_k = 0 \,. \tag{30}$$

We say that $e, e_1, e_2, \ldots$ form a chain of joined vectors. The maximal order m of vectors, joined to e is finite and the number $m + 1$ is called multiplicity of the eigenvector e.

Definition 3.2. *The set of eigenvectors and joined vectors*

$$e^{(k)}(-\lambda_j), e_1^{(k)}(-\lambda_j), \ldots, e_{m_k}^{(k)}(-\lambda_j) \qquad (k = 1, 2, \ldots, N(-\lambda_j)) \tag{31}$$

corresponding to an eigenvalue $-\lambda_j$ is called canonical system if:

i) *Vectors* $e^{(k)}(-\lambda_j), k = 1, 2, \ldots, N(-\lambda_j)$ *form a basis in the space of eigenvectors corresponding to the eigenvalue* $-\lambda_j$.

ii) $e^{(1)}(-\lambda_j)$ *is an eigenvector with maximal possible multiplicity.*

iii) $e^{(k)}(-\lambda_j)$ *is an eigenvector which can not be expressed by a linear combination of* $e^{(1)}(-\lambda_j), \ldots, e^{(k-1)}(-\lambda_j)$ *and multiplicity of* $e^{(k)}(-\lambda_j)$ *achives a possible maximum.*

iv) *Vectors (31) with fixed k form a complete chain of joined elements.*

Besides canonical system (31) which corresponds to an eigenvalue $-\lambda_j$ of operator $-A$ we consider a canonical system

$$\varepsilon^{(k)}(-\overline{\lambda}_j), \varepsilon_1^{(k)}(-\overline{\lambda}_j), \ldots, \varepsilon_{m_k}^{(k)}(-\overline{\lambda}_j) \qquad (k = 1, 2, \ldots, N(-\overline{\lambda}_j)) \tag{32}$$

that corresponds to the eigenvalue $-\overline{\lambda}_j$ of the ajoint operator $-A^*$. Definition of canonical system (32) is absolutely analogous to Definition 3.2 of canonical system (31). We define canonical system (32) by $E^*(-\overline{\lambda}_j)$.

Theorem 3.1 *Let* R_k *are operators defined in (27). Then*

$$R_{-k}x = 0, \quad \forall k = 1, 2, \ldots, m \tag{33}$$

if and only if

$$\langle x, \varepsilon_l^{(k)}(-\overline{\lambda}_j)\rangle = 0 \qquad \forall \varepsilon_l^{(k)}(-\overline{\lambda}_j) \in E^*(-\overline{\lambda_j})\,. \tag{34}$$

This assertion follows immediately from one result of Keldysh ([K]) on structure of the main part of Laurent serie for $R(\lambda, -A)$. The proof of Theorem 3.1 see in [F3].

We write boundary value problem (22), (23) for Ozeen equations written in the form

$$\frac{dw(t)}{dt} + Aw(t) = 0, \quad w|_{t=0} = w_0 \tag{35}$$

where A is operator (26). Then for each $w_0 \in V_0^0(G)$ the solution $w(t,\,\cdot)$ of (35) is defined by $w(t,\,\cdot) = e^{-At}w_0$ and

$$e^{-At} = (2\pi i)^{-1}\int_\gamma (\lambda I + A)^{-1}e^{\lambda t}d\lambda\,, \tag{36}$$

where γ is a contur belonging to $\rho(-A)$ such that $\arg\lambda = \pm\theta$ for $\lambda \in \gamma, |\lambda| \geqslant N$ for certain $\theta \in (\pi/2, \pi)$ and for sufficiently large N. Moreover, γ surrounds $\Sigma(-A)$ from the right. Such contur γ exists, of course, because A is a sectorial operator.

Let $\sigma > 0$ satisfy:

$$\Sigma(-A) \cap \{\lambda \in \mathbb{C} : \mathrm{Re}\lambda = -\sigma\} = \emptyset\,. \tag{37}$$

The case when there are certain points of $\Sigma(-A)$ placed righter than the line $\{\mathrm{Re}\lambda = -\sigma\}$ will be interesting for us. By γ_σ we denote the continuous contur that is placed in $\{\lambda \in \mathbb{C} : \mathrm{Re}\lambda \leqslant -\sigma\}$ and constructed from an interval of the line $\{\mathrm{Re}\lambda = -\sigma\}$ and from two branches of contour γ that transform to $\{\arg\lambda = \theta\}$ and $\{\arg\lambda = -\theta\}$, $\theta \in (\pi/2, \pi)$ for sufficiently large $|\lambda|$.

We reduce integration over γ in (36) to integration over γ_σ and integration around poles $-\lambda_j$ from (27) for λ_j satisfying $\mathrm{Re}\lambda_j < \sigma$ After calculation corresponding residues we transform (36) to the equality:

$$e^{-At} = \int_{\gamma_{\sigma_0}} (\lambda E + A)^{-1} \frac{e^{\lambda t} d\lambda}{2\pi i} + \sum_{\mathrm{Re}\lambda_j < \sigma} e^{-\lambda_j t} \sum_{n=1}^{m(-\lambda_j)} \frac{t^{n-1} R_{-n}(-\lambda_j)}{(n-1)!} . \tag{38}$$

We denote $V_0^1(G) = \{v \in V^1(G) : v|_{\partial G} = 0\}$.

Theorem 3.2. *Suppose that A is operator* (26) *and $\sigma > 0$ satisfies* (37). *Then for each $w_0 \in V_0^1(G)$ that satisfies*

$$\langle w_0, \varepsilon_l^{(k)}(-\bar{\lambda}_j) >= 0, \quad l = 0, 1, \ldots, m_k, \ k = 1, 2, \ldots, N(-\lambda_j) , \tag{39}$$

for any λ_j with $\mathrm{Re}(\lambda_j) < \sigma$ the following inequality holds: [1]

$$\|e^{At} w_0\|_{V_0^1(G)} \leqslant c e^{-\sigma t} \|w_0\|_{V_0^1(G)} \quad \textit{for } t \geqslant 0 . \tag{40}$$

Proof. By Theorem 3.1 if w_0 satisfies (39) then $R_n(-\lambda_j) w_0 = 0$ for every operators $R_{-n}(-\lambda_j)$ from (38). One can prove easily that

$$\|(\lambda I + A)^{-1} v\|_{V_0^1(G)} \leqslant c \|v\|_{V_0^1(G)} \cdot$$

with $c > 0$ which does not depend on $v \in V_0^1(G), \lambda \in \gamma_\sigma$. Therefore (38) implies (40).

Impose on canonical systems (32) the following condition

$$\varepsilon^{(k)}(-\lambda_j) = \bar{\varepsilon}^{(k)}(-\bar{\lambda}_j); \quad \varepsilon_l^{(k)}(-\bar{\lambda}_j) = \bar{\varepsilon}_l^k(-\lambda_j) \tag{41}$$

Condition (41) can be realized by (29).

In virtue of (41) canonical system corresponding real $-\lambda_j$ consists of real valued vector fields. If $\mathrm{Im}\lambda_j \neq 0$, instead of vector fields $\varepsilon^{(k)}(-\bar{\lambda}_j)$, $\bar{\varepsilon}_l^{(k)}(-\lambda_j)$, $l = 0, 1, \ldots,$ we consider real valued vector fields

$$\mathrm{Re}\varepsilon_l^{(k)}(-\bar{\lambda}_j), \mathrm{Im}\varepsilon_l^{(k)}(-\bar{\lambda}_j), \quad l = 1, \ldots, \quad k = 1, 2 \ldots . \tag{42}$$

[1]Here by definition $\varepsilon_0^{(k)}(-\lambda_j) = \varepsilon^{(k)}(-\lambda_j))$

We renumber all functions (42) with $\mathrm{Re}\lambda_j < \sigma$ (including fields with $\mathrm{Im}\lambda_j = 0$) as follows:

$$\varepsilon_1(x), \ldots, \varepsilon_K(x)\,. \tag{43}$$

Corollary 3.1. *Assume, that A is operator (26) and $\sigma > 0$ satisfies (37). Then for each $w_0 \in V_0^1(G)$ satisfying*

$$\int_G (w_0(x), \varepsilon_j(x))\partial x = 0, \qquad j = 1, .., K \tag{44}$$

with ε_j from (43), inequality (40) is true.

4 Stabilization of the Ozeen Equation

4.1 Theorem on Extension

The key step in stabilization method that we propose is construction of special extension for initial vector field $v_0(x)$ from (3.15) defined on Ω to a vector field defined on G which we take as initial value $w_0(x)$ in (3.23). First of all we make more precise the conditions imposed on Ω and G. Recall that by (3.20), (3.16)

$$G = \mathrm{Int}(\bar{\Omega} \cup \bar{\omega}), \quad \partial\Omega = \Gamma \cup \Gamma_0 \cup \partial\Gamma, \quad \partial G \cap \partial\Omega = \Gamma_0 \cup \partial\Gamma \tag{45}$$

where Ω and ω are open subsets of $\mathbb{R}^2$, $\Omega \cap \omega = \emptyset$, $\partial\Omega$ is a closed C^∞-curve, and Γ, Γ_0 are open subsets of $\partial\Omega$, $\Gamma \neq \emptyset$, $\partial\Gamma$ is a finite number of points, or $\partial\Gamma = \emptyset$. We suppose that $\partial\Omega = \bigcup_{j=1}^N \partial\Omega_j$ where $\partial\Omega_j$ are connected components of $\partial\Omega$. We impose

Condition 4.1. *For each $j = 1, \ldots, N$ the set $\partial\Omega_j \cap \Gamma_0$ is connected or empty.*

Condition 4.2. *Let $\partial\Omega \in C^\infty$, $\partial G \setminus \partial\Gamma \in C^\infty$ and for each point $P \in \partial\Gamma$ there exist local coordinates (x, y) such that P is origin; $P = (0,0)$, $\{(x,0), x \in (0,\varepsilon)\} \subset \Gamma$, $\{(x,0),\ x \in (-\varepsilon, 0)\} \subset \Gamma_0$ where $\varepsilon > 0$ is small enough and*

$$\partial G \supset \{(x,y) = (x, x^\alpha),\ x \in (0,\varepsilon)\} \cup \{(x,0),\ x \in (-\varepsilon,0)\},\ \alpha \geqslant 2\,. \tag{46}$$

Since Ω is a given domain where stabilized Ozeen system is determined, and we choose ourselves the domain ω, Condition 4.2 is not restrictive.

We introduce the following space

$$V^1(\Omega, \Gamma_0) = \{u(x) \in V^1(\Omega) : u|_{\partial\Gamma_0} = 0,\ \exists v \in V_0^1(G) \text{ that } u = \gamma_\Omega v\} \tag{47}$$

where γ_Ω is the operator of restriction on Ω for functions from $V_0^1(G)$.

We formulate now the extension theorem. In the space of real valued vector fields $V_0^1(G)$ we introduce the subspace

$$X_\sigma^1(G) = \{v(x) \in V_0^1(G) : \int_G v(x) \cdot \varepsilon_j(x)\, dx = 0, \quad j = 1, \ldots, K\} \tag{48}$$

where $\varepsilon_j(x)$ are functions (3.43).

Theorem 4.1. *There exists a linear bounded extension operator*

$$E_\sigma^1 : V^1(\Omega, \Gamma_0) \to X_K^1(G) \tag{49}$$

(i.e. $E_\sigma(v)(x) \equiv v(x)$ for $x \in \Omega$).

Proof of this Theorem see in [F3].

We show the conditions on $v \in V^1(G)$ which garantee inclusion $v \in V^1(\Omega, \Gamma_0)$

Theorem 4.2. *Let domains Ω and G satisfy* (45), *Conditions* 4.1, 4.2 *and $\partial\Gamma \neq \emptyset$. Suppose that $v(x) \in V^1(\Omega)$ satisfies the conditions: $v|_{\Gamma_0} = 0$ and*

$$v(x) \in \left(H^m(\Omega \cap \mathcal{O}(\partial\Gamma))\right)^2 \text{ with } m > \frac{3\alpha - 1}{2}, \ \alpha \geqslant 2\,, \tag{50}$$

where $\mathcal{O}(\partial\Gamma)$ is a neighbourhood of $\partial\Gamma$ and α is a magnitude from (46). *Then $v(x) \in V^1(\Omega, \Gamma_0)$, i.e. it can be extended up to a vector field $Lv \in V_0^1(G)$.*

Proof see in [F3].

4.2 Result on Stabilization

We establish now the main theorem of this section on stabilizability by feedback boundary control of 2D Ozeen equations.

Theorem 4.3. *Let domains Ω and G satisfy* (45), *and Conditions* 4.1, 4.2. *Then for each initial condition $v_0(x) \in V^1(\Omega, \Gamma_0)$ and for each $\sigma > 0$ there exists a feedback control u defined on Σ such that the solution $v(t,x)$ of* (3.13)–(3.15), (3.17) *satisfies the inequality*

$$\|v(t, \cdot)\|_{(H^1(\Omega))^2} \leqslant ce^{-\sigma t} \quad \text{as} \quad t \to \infty\,. \tag{51}$$

Proof of this Theorem is based on Theorem 4.1 (see [F3]).

4.3 Feedback Property

We outline how the method of stabization proposed above can be used for construction of a control which reacts on unpredictable fluctuations of a system. The point is that if the solution $(v(t,x), u(t,x))$ of problem (3.13)–(3.15),(3.17) satisfies equalities (3.25) and if at time moment $\tilde{t}_0$ the system (3.13)–(3.15) is subjected by certain fluctuation, then $v(t,x)$ at $t = \tilde{t}_0$ is pushed out $\gamma_\Omega X_\sigma^1(G)$ and that is why it will not tend to zero with prescribed rate. Therefore we check when

$$\|v(t,\cdot) - \gamma_\Omega w(t,\cdot)\|_{V_0^1(\Omega)} \geqslant ce^{-\sigma t/2} \tag{52}$$

(here $w(t,x)$ is the solution of (3.35) with $w_0(x) = E^1_\sigma v_0(x)$) and at this moment, say t_1, we regard $v(t_1,x)$ as initial condition, and for $t > t_1$ we construct the solution (v,u) of (3.13)–(3.15),(3.17) by formulas (3.25) where $w(t,x)$ is the solution of (3.23_1) with initial condition

$$w|_{t=t_1} = E^1_\sigma v(t_1,\cdot)$$

This construction can be written breafly as umpulse control for (3.35), concentrated in the artificial part $\omega = G \setminus \Omega$ of domain:

$$\partial_t w(t,x) + Aw = \delta(t-t_1)(E^1_K v(t_1,x) - \tilde{w}(t_1,x))$$

where $\tilde{w}(t_1,x) = w(t_1,x)$, for $x \in \omega$ and $\tilde{w}(t_1,x) = v(t_1,x)$, for $x \in \Omega$ After next unpredictable pushing out $\gamma_\Omega X^1_K(G)$ of the solution $v(t,x)$ we do the same.

5 Stabilization of 2D Navier–Stokes Equations

5.1 Formulation of the Stabilization Problem

We consider the 2D Navier–Stokes equations

$$\partial_t v(t,x) - \Delta v(t,x) + (v,\nabla)v + \nabla p(t,x) = f(x), \quad \operatorname{div} v = 0 \tag{53}$$

$$v(t,x)|_{t=0} = v_0(x), \qquad v|_{\Sigma_0} = 0, \quad v|_\Sigma = u\,, \tag{54}$$

where $u = (u_1,u_2)$ is a control defined on Σ. We suppose also that a steady-state solution $(\hat{v}(x), \nabla\hat{p}(x))$ of Navier–Stokes system with the same right-hand side $f(x)$ as in (53) is given:

$$\Delta\hat{v}(x) + (\hat{v},\nabla)\hat{v} + \nabla\hat{p} = f(x), \quad \operatorname{div}\hat{v}(x) = 0, \quad \hat{v}|_{\Gamma_0} = 0\,. \tag{55}$$

Let $\sigma > 0$ be given. The problem of stabilization with the rate σ is to look for a solution $(v,\nabla p,u)$ of problem (53), (54) which satisfies the inequality

$$\|v(t,\cdot) - \hat{v}(\cdot)\|_{(H^1(\Omega))^2} \leqslant ce^{-\sigma t} \quad \text{as } t \to \infty\,. \tag{56}$$

The important additional condition which we impose is that u is a feedback control. We give now exact formulation of feedback control although it is analogous to Definition 3.1.

We extend Ω to a domain G throught Γ (see (4.45) such that Condition 4.2 holds. After that we extend problem (53), (54) to analogous problem defined on $\Theta = \mathbb{R}_+ \times G$ with zero boundary condition on $S = \mathbb{R}_+ \times \partial G$:

$$\partial_t w(t,x) - \Delta w + (w,\nabla)w + \nabla q(t,x) = g(x), \quad \operatorname{div} w(t,x) = 0\,, \tag{57}$$

$$w(t,x)|_{t=0} = w_0(x), \qquad w|_S = 0\,. \tag{58}$$

Moreover, we assume that solution $(\hat{v}, \nabla\hat{p})$ of (55) is extended on G to a pair $(a(x), \nabla\hat{q}(x))$, $x \in G$ such that

$$-\Delta a(x) + (a,\nabla)a + \nabla\hat{q}(x) = g(x), \quad \operatorname{div} a(x) = 0, \quad a|_{\partial G} = 0 \tag{59}$$

where right side $g(x)$ is the same as in (57).

Definition 5.1 *A control $u(t,x)$ in stabilization problem (53), (54), (56) is called feedback if the solution $(v(t,x), u(t,x))$ of (53), (54) is defined by the equality:*

$$(v(t,x), u(t,x)) = (\gamma_\Omega w(t,\,\cdot\,), \gamma_\Gamma w(t,\,\cdot\,)) \tag{60}$$

where $w(t,x)$ satisfies to (57), (58), and γ_Ω, γ_Γ are restriction operators (3.24).

5.2 Invariant Manifolds

Let $g(x)$ from (57) satisfies: $g(x) \in (L_2(G))^2$. Then, as well-known, (57) are equivalent to the following equation with respect to one unknown function $w(t,x)$:

$$\partial_t w(t,x) - \pi\Delta w + \pi(w,\nabla)w = \pi g(x) \tag{61}$$

where π is orthoprojector on $V_0^0(G)$. We assume in addition that solution w of (61) (as well as soulution w of (57) belongs to the space

$$V_0^{1,2}(\Theta_T) \equiv L_2(0,T; V^2(G) \cap (H_0^1(G))^2) \cap H^1(0,T; V_0^0(G)) \tag{62}$$

for each $T > 0$, where $\Theta_T = (0,T)\times G$. It is proved that for each $T > 0$, $g(x) \in (L_2^0(G))^2$, $w_0(x) \in V_0^1(G)$ there exists unique solution $w(t,x) \in V_0^{1,2}(Q_T)$ of problem (61), (58). Solution $w(t,x)$ of (61), (58) taken at time moment t we denote as $S(t,w_0)(x) :\; w(t,x) = S(t,w_0)(x)$.

Since embedding $V_0^{1,2}(Q_T) \subset C(0,T; V_0^1(G))$ is continuous, the family of operators $S(t,w_0)$ is continuous semigroup defined on the space $V_0^1(G)$: $S(t+\tau, w_0) = S(t, S(\tau, w_0))$.

Note that we can rewrite (59) in the form analogous to (61):

$$-\pi\Delta a(x) + \pi(a,\nabla)a = \pi g, \quad a(x) \in V^2(G) \cap V_0^1(G)\,. \tag{63}$$

Since $a(x)$ is steady-state solution of (61), $S(t,a) = a$ for each $t \geqslant 0$. We can decompose semigroup $S(t,w_0)$ in a neighbourhood of a in the form

$$S(t, w_0 + a) = a + L_t w_0 + B(t, w_0) \tag{64}$$

where $L_t w_0 = S'_w(t,a)w_0$ is derivative of $S(t,w_0)$ with respect to w_0 at point a, and $B(t,w_0)$ is nonlinear operator with respect to w_0. Differentiability of $S(t,w_0)$ is proved, for instance in [BV, Chap. 7. Sect. 5]. Therefore

$$B(t,0)=0, \quad B'_w(t,0)=0\,. \tag{65}$$

Moreover in [BV, Chap. 7. Sect. 5] is proved that $B(t,w)$ belongs to class C^α with $\alpha=1/2$ with respect to w.

We study now semigroup $L_t w_0 = S'_w(t,a)w_0$ of linear operators. First of all note that $w(t,x)=L_t w_0$ is the soulution of problem (3.22)–(3.23) in which the coefficient a is the soulution of (63). Therefore $L_t w_0 = e^{-At}w_0$ where A is Ozeen operator (3.26).

Below we suppose that $r_0\in(0,1)$ satisfies the property:

$$\{\zeta\in\mathbb{C}\,:\,|\zeta|=r_0\}\cap\Sigma(e^{-At_0})=\emptyset \tag{66}$$

where, recall, $\Sigma(e^{-At})$ is the spectrum of operator $L_t w_0 = e^{-At}w_0$.

It is clear, that $\zeta_j\in\Sigma(e^{-At_0})$ if and only if $\zeta_j=e^{-\lambda_j t_0}$ and $-\lambda_j\in\Sigma(-A)$. That is why condition (66) is equivalent to condition (3.37) where $\sigma=-\ln r_0/t_0$. Besides, if $|\zeta_j|>r_0$ then $-\mathrm{Re}\lambda_j>-\sigma$.

Theorem 5.1. *Family of operators* $e^{-At}:V_0^1(G)\to V_0^1(G)$ *where* A *is operator* (3.26) *is well defined for each* $t\geqslant 0$. *Let*

$$\sigma_+=\{\zeta_1,\dots,\zeta_N\,:\,\zeta_j\in\Sigma(e^{-At_0}),\ \ |\zeta_j|>r_0,\ j=1,\dots,N\} \tag{67}$$

where $r_0\in(0,1)$ *and satisfies* (66). *Let* $X_+\subset V_0^1(G)$ *be the invariant subspace for* e^{-At_0} *corresponding to* σ_+, $\Pi_+:V_0^1(G)\to X_+$ *be the projector on* X_+ *(i.e.,* $\Pi_+V_0^1(G)=X_+$, $\Pi_+^2=\Pi$*) and* $X_-=(I-\Pi_+)V_0^1(G)$ *be complementary invariant subspace. Let* $L_{t_0}^+=e^{-At_0}|_{X_+}:X_+\to X_+$, $L_{t_0}^-=e^{-At_0}|_{X_-}:X_-\to X_-$. *Then operator* $L_{t_0}^+$ *has inverse operator* $(L_{t_0}^+)^{-1}$. *For some* t_0 *there exist constants* $\hat r$, ε_+, $\varepsilon_-\in(0,1)$ *such that*

$$\|L_{t_0}^-\|\leqslant\hat r(1-\varepsilon_-), \quad \|(L_{t_0}^+)^{-1}\|\leqslant\hat r^{-1}(1-\varepsilon_+)\,. \tag{68}$$

Generally speaking eigenvalues of operators A and e^{-At} are complex-valued. That is why all spaces in Theorem 5.1 are complex. But to apply obtained results to (nonlinear) Navier–Stokes equation we need to have analogous results for the real spaces of the same type. Actually, for this we have to define the projector of Π_+ in real spaces.

Lemma 5.1. *Restriction of operator* Π_+ *on the real space* $V_0^1(G)$ *can be written in the form*

$$(\Pi_+)(x)=\sum_{j=1}^{K}e_j(x)\int_G v(x)\varepsilon_j(x)\,dx \tag{69}$$

where $\{\varepsilon_j\}$ *is the set of functions (3.43) which are suitably renumbered and renornalized functions (3.42) and* $\{e_j\}$ *is set of Real and Imaginary parts of functions (3.31) analogously renumbered and renornalized.*

Using (69) we can easily restrict spaces X_+ and X_- as well as operators $L^+_{t_0}$, $L^-_{t_0}$ defined in formulation of Theorem 5.1 on the real subspaces of $V^1_0(G)$. We denote this new real spaces and operators also by X_+, X_-, $L^+_{t_0}$, $L^-_{t_0}$. This will not lead to misunderstanding because below we do not use their complex analogs.

In a neighbourhood of steady-state solution a of (63) we establish existence of a manifold M_- which is invariant with respect to semigroup $S(t,w)$ (i.e., $S(t,w) \in M_- \ \forall w \in M_- \forall t > 0$). This manifold can be represented as the graph:

$$M_- = \{u \in V^1_0(G) : u = a + u_- + g(u_-) , \quad u \in X_- \cap \mathcal{O}\} \tag{70}$$

where $\mathcal{O}$ is a neighbourhood of origin in $V^1_0(G)$, $g : X_- \cap \mathcal{O} \to X_+$ is an operator-function of class $C^{3/2}$ and

$$g(0) = 0, \quad g'(0) = 0 \, . \tag{71}$$

Note that this condition means that manifold (70) is tangent to X_- at point a.

Theorem 5.2. *Let a satisfy* (63), $\sigma > 0$ *satisfy* (3.37), $\mathcal{O} = \mathcal{O}_\varepsilon = \{v \in V^1_0(G) : \|v\|_{V^1_0(G)} < \varepsilon\}$ *and ε is sufficiently small. Then there exists unique operator-function $g : X_- \cap \mathcal{O} \to X_+$ of class $C^{3/2}$ satisfying* (71) *such that the manifold M_- defined in* (70) *is invariant with respect to semigroup $S(t,w_0)$ connected with* (61)[2]. *There exists a constant $c > 0$ such that*

$$\|S(t,w_0) - a\|_{V^1_0(G)} \leqslant c\|w_0 - a\|_{V^1_0(G)} e^{-\sigma t} \quad \text{as } t \geqslant 0 \tag{72}$$

for each $w_0 \in M_-$.

This theorem follows form results of [BV, Chap. 5, Sect. 2; Chap. 7, Sect. 5] and from Theorem 5.1.

5.3 Final Results

Here we construct extension operator for Navier–Stokes equations. This operator is nonlinear analog of extension operator (4.49) constructed for Ozeen equatios.

Theorem 5.3. *Suppose that $a(x)$ is a steady-state solution of* (63), $\hat{v}(x) = \gamma_\Omega a$, *and M_- is the invariant manifold constructed in a neighbourhood $a + \mathcal{O}$ of a in $V^1_0(G)$ in Theorem* 5.2. *Let*

$$B_{\varepsilon_1} = \{v_0 \in V^1(\Omega, \Gamma_0) : \|v_0 - \hat{v}\|_{V^1(\Omega,\Gamma_0)} < \varepsilon_1\} \, .$$

Then for sufficiently small ε_1 one can construct a continuous operator

$$\mathrm{Ext} : \hat{v} + B_{\varepsilon_1} \to M_- \, , \tag{73}$$

[2] I.e. $S(t,w_0)$ is the resolving semigroup of equation (61).

which is operator of extension for vector fields from Ω to G:

$$(\text{Ext}v)(x) \equiv v(x), \quad x \in \Omega . \tag{74}$$

The proof of this theorem see in [F3].

We set

$$V^2(\Omega, \Gamma_0)$$
$$= \{v(x) \in V^2(\Omega) : v|_{\Gamma_0} = 0, \ \exists w \in V^2(G) \cap V_0^1(G), \ v(x) = \gamma_\Omega w\} \tag{75}$$

Proposition 5.1. *Let $f \in (L_2(\Omega))^2$ and $(\hat{v}(x), \nabla\hat{p}(x)) \in V^2(\Omega, \Gamma) \times (L_2(\Omega))^2$ satisfies equations* (55). *Then there exist an extension $g(x) \in (L_2(G))^2$ of $f(x)$ from Ω to G and an extension $(a(x), \nabla q(x)) \in (V^2(G) \cap V_0^1(G)) \times (L_2(G))^2$ of $(\hat{v}(x), \nabla\hat{p}(x))$ from Ω to G such that the pair $(a(x), \nabla q(x))$ is a solution of* (59).

We now are in position to formulate the main result of this paper.

Theorem 5.4. *Let $\Omega \subset \mathbb{R}^2$ be a bounded domain with C^∞-boundary $\partial\Omega$ and $\partial\Omega = \Gamma_0 \cup \Gamma \cup \partial\Gamma$, where Γ, Γ_0 are open curves, $\Gamma \neq \emptyset$, $\partial\Gamma$ is a finite number of points or $\partial\Gamma = \emptyset$. Suppose that a domain $G \subset \mathbb{R}^2$ is chosen such that assumption* (4.45) *and Conditions* 4.1, 4.2 *with $\alpha \geqslant 3$ are fulfilled. Let $f(x) \in (L_2(\Omega))^2$ and $(\nabla\hat{v}(x), \nabla\hat{p}(x)) \in V^2(\Omega, \Gamma_0) \times (L_2(\Omega))^2$ satisfy* (55). *Then for an arbitrary $\sigma > 0$ there exists sufficiently small $\varepsilon > 0$ such that for each $v_0 \in V^1(\Omega, \Gamma_0)$ satisfying $\|\hat{v} - v_0\|_{V_0^1(\Omega)} < \varepsilon_1$ there exists a feedback boundary control $u(t,x)$, $(t,x) \in \Sigma \equiv \mathbb{R}_+ \times \Gamma$ which stabilized Navier–Stokes boundary value problem* (53)–(54) *with the rate* (56), *i.e. the soulution v of* (53)–(54) *satisfies* (56).

The proof of this theorem is based on Theorems 5.2, 5.3 (see [F2], [F3]).

References

[BK] Balogh A., Krstic M. (2001) Regularity of solutions of Burgers' Equation with globaly Stabilizing Nonlinear boundary Feedback SIAM J. Control and Optimization (to appear).

[BV] Babin A.V., Vishik M.I. (1992) Attractors of evolution equations. North-Holland, Amsterdam, London, New York, Tokio.

[C] Coron J.-M. (2001) On null asymptotic stabilization of the 2-D Euler equation of incompressible fluids on simply connected domains. (to appear)

[F1] Fursikov A.V. (2000), Optimal control of distributed systems. Theory and applications. Translations of Math. Monographs, **187** Amer. Math. Soc., Providence, Rhode Island, 2000.

[F2] Fursikov A.V. (2001) Stabilizability of quazilinear parabolic equation by feedback boundary control. Matem. Sbornik, (to appear)

[F3] Fursikov A.V. (2001) Stabilizability of two-dimensional Navier–Stokes equations with help of boundary feedback control, J. of Math. Fluid Mechanics, (to appear)

[FE] Fursikov A.V., Emanuilov O. Yu. (1999) Exact controllability of Navier–Stokes and Boussinesq equations, Russian Math. surveys, 54:3: 565–618.

[K] Keldysh M.V. (1971) On completeness of eigen functions for certain classes of not self-adjoint linear operators. Russian Math Surveys, 26:4: 15–41, (in Russian).

[Lag] Lagnese J. (1989) Boundary stabilization of thin plates. SIAM, Phyladelphia

[Li] Lions J.-L. (1988) Exact controllability, stabilizability, and perturbations for distributed systems. SIAM Rev 30 1–68.

Vortex-based Control Algorithms

Dmitri Vainchtein and Igor Mezić

Department of Mechanical Engineering, University of California Santa Barbara, CA 93106, USA

Abstract. In high Reynolds number flows, interactions between coherent vortical structures are key to understanding dynamics. In this paper we review methods of control that rely on this observation. Control of vortex dynamics is pursued using a variety of reduced-order representations of the dynamics, such as point vortices, vortex blobs and vortex patches. Control algorithms designed based on such representations are applied in many contexts of which we review a few of the most common ones, such as wake vorticity control and recirculation zone vortex control. We also review some of our own work in the area of vortex control concentrated on two-vortex merging interactions. The methods used are those of averaging in the case of limited control authority and the nonlinear control method of flat coordinates.

1 Introduction

In realistic flows, the presence of multiple global (possibly unstable) modes necessitates the use of multiple control sensors at many locations for the identification, monitoring and (if needed) suppression of all possible unstable modes. The spatio-temporal response of a full hydrodynamical structure is nonlinear. Consequently, knowledge of flow variables (for example velocity or pressure) at many points throughout the flow region is needed within the feedback control algorithm. Therefore, the resulting control algorithm may (and probably will) be complicated and computationally slow, although the advantage of obtaining very accurate solutions may outweigh those disadvantages. For examples of such an approach (implemented to the problem of the vortex/wall interaction) see e.g. [35, 36]. In certain cases, however, a complex distributed behavior can be addressed locally, as a complex spatiotemporal behavior is governed by (relatively) few modes. Some theoretical results on possibility of localized control of dynamical systems possessing spatiotemporal chaos was presented in [7]. In that case, a low dimensional description of the flow features and their response to external forcing is possible and feedback control algorithms can be made simpler and computationally faster (see e.g. [23, 34]).

In most of the problems, for analytical or numeric solution to be achieved, certain simplifications must be made. There are two strategies for such simplifications. Namely, one can simplify the *equations* or simplify the *model* (for example, by reducing the global dynamics to the interaction of coherent structures). In the following we concentrate on the model-based approaches. Concentrated vortices are a natural candidate to be used for model-based description.

In this paper, we first present a short review of efforts in control of vortex dynamics and then present summary of our own efforts in this area, introducing methods for control of vortex elements based on averaging and the concept of flat coordinates.

2 Review of the Vortex Control Research

In the present section we discuss the following topics of the control of vortex dynamics:

(i) Control of a set of interacting N (typically for small N) vortices (point vortices or vortex blobs)
(ii) Trapping of vortices near a body.
(iii) Vorticity control for robotic fish swimming.
(iv) Control of mixing.
(v) Control of the wake behind a bluff body.

With the above list we did not try to cover the entire scope of the vortex control problems. Instead, we our aim was to illustrate the spectrum of applicability of the vortex based control algorithms, from simple configurations of a few point vortices to the complex structure of the wakes. Work on control of vortex dynamics can be grouped into two categories. In one case (topics 1,2 above) the objective is to make vortices move along a prescribed path (for example, stabilize an unstable equilibrium), in the other the main objective is mixing, prevention of vortex shedding or enhancement of vortex merger (topics 3,4 and 5) in which the exact trajectory of vortex elements is the consequence of that objective.

2.1 Simple Vortical Configurations

An important (and probably the simplest) setting of the vortex dynamics involves a few (point) vortices in a relatively simple domain. In this case methods of dynamical systems theory are readily applicable and can be used to manipulate the vortices into any prescribed path. In such problems, the knowledge of the model equations is assumed and the number of intrinsic parameters is typically small. Consequently, much of the work can be done analytically.

Additional simplification comes from the Hamiltonian structure of the problem of 2-D point vortices (see e.g. [8, 13, 70]), or, more generally for compact vortex clusters, where the state of a cluster can be described by just a few variables). This property becomes especially important for systems with few vortices where Hamiltonian structure yields additional integrals of motion and can be used to describe the structure of the phase space (for general discussion of control of Hamiltonian systems, see e.g. [71]).

Research on control of a set of point vortices was pioneered by Pentek and co-workers, [77], who applied a standard Ott-Grebogy-Yorke (OGY) technique to actively stabilize unstable periodic orbits of point vortices inside a circular cylinder wall (with actuators being a set of sinks at the wall). Friedland and Shagalov [30, 31] discussed theoretically and numerically producing and controlling so-called V-shape vortices using external strain fields. Friedland, [29], discussed theoretically and numerically resonance control of Kirchhoff vortices (2-D elliptical vortex patches) using an autoresonance between co-rotation of the vortices and time dependent strain field. A similar configuration was considered numerically in [19] where a possibility of vortex merging was included.

2.2 Vortex/Solid Body Interaction

Major problems of interest for control of vortex interaction are the vortex/wall interaction and wakes behind bodies. Airfoils provide one of the most important and well-known examples of such interactions. It was shown that a vortex trapped at the airfoil improves the lift and reduces drag (see e.g. [55]). Most of configurations of vortices near the wall in the presence of the external flow were found to be unstable. Left to their own devices, vortices tend either to be swept away by the external flow and/or start shedding secondary vortices. Either process may be undesirable. Therefore, a variety of mechanical and/or control techniques were used to enforce trapping of vortices. The research on control of the vortex/wall interaction and vortex shedding utilises a wide range of methods from point vortices to vortex blob methods to the direct analysis of Navier-Stokes equations. A common property of the all the papers cited below is that the major objective is to create and/or capture a vortex and keep it relatively intact.

A basic configuration used to study the vortex/body interaction is the one consisting of a point vortex and a cylinder (various geometries can be reduced to the cylinder using a conformal mapping). Chernyshenko, [21], provided an analytical algorithm that uses sources on the surface of a cylinder to keep a point vortex near it. Pentek and co-workers, [53, 76], used a modified OGY chaos control scheme, based on a low-dimensional Hamiltonian model of the flow, to capture and stabilize a concentrated vortex in the vicinity of a cylinder. The actuation is done by uniformly rotating the cylinder and changing the background flow velocity far from the cylinder. The suggested control mechanism, when extended to non-point vortices, achieves two goals: to control the dynamics of the vortex and to suppress the vortex shedding. In [76], the OGY method was used for control of a chaotic vortex/body interaction.

In addition to modifying the flow around the body, one can modify the body itself to improve the trapping. Iollo and Zanetti [50, 51] discussed controlling the motion of a point vortex trapped in an artificial cavity at the upper side of an airfoil subject to the potential external flow. The main objective is to return the vortex to the equilibrium position if the vortex is displaced by external perturbations and thus improve the airfoil lift. The actuators are sources/sinks at the wall and the suction/blowing law is obtained by the adjoint optimization method.

Analytical approach implementing models more complex than point vortices started with the papers of Cortelezzi and co-workers, [23, 24, 25], who considered the problem of maintaining a vortex near a corner of a wall. Most of the work in this area is concentrated on preventing vortex shedding. A standard actuation is one or more sink/sources at the wall. In [24], the authors discussed a hierarchy of models of increasing complexity: from a low-order point vortex model to a high-order vortex blob model. A nonlinear controller capable of manipulating the wake by means of a suction point located on the downstream wall of the plate is given in closed form for the point vortex model. It was shown that a controller devised based on a simpler (e.g. point vortex model) is doing a good job for more advanced (e.g. vortex blob) model. This approach became an important tool in vortex-based research. Iollo and Zanetti, [97], applied it to devise analytically a passive control, based on the wall suction acting at the leading edge, to stabilize the vortex shedding from a flat plate at incidence. They used the algorithm in a vortex blob simulation to check the stabilization of the unsteady wake past the plate. Anderson et al., [5], used the vortex-blob method to construct numerically a PI controller used to prevent the shedding.

A review of certain control mechanisms for a full system of the interaction between a body and incidenting vortex was presented in [80]. The case of unsteady compressible vortical fluids was discussed in [22] where a general framework of adjoint approach to optimal control such systems was introduced and applied to vortex/wall interaction.

2.3 Control of the Wake Behind a Bluff Body

Among all the configurations possessing the concentrated vorticity, the wake behind a body is probably the most well known. It is well understood that changing the vortical structure of the wake can be both an objective of the control algorithm and/or a useful method for achieving other objectives, like modifying the pressure distribution. Different approaches for the control of a flow around a body (e.g. a cylinder) have been successfully employed in the last two decades. Modi, [69], conducted experiments on the role of rotating cylinder(s) on the wake structure near airfoil. Tang and Aubry, [84], studied numerical suppressing of the vortex shedding by inserting two small vortex perturbations in the flow; Gillies, [34], used POD to obtain a lower-order model that he solved partially analytically, partially numerically using a neural networks method. Kwon and Choi, [57], (numerically) and Ozono, [74] (experimentally) studied the impact of splitter plates attached to the cylinder. Feedback control was studied numerically by Park and co-workers, [75], who used a pair of blowing/suction slots located on the surface of the cylinder; by Tao and co-workers, [85], who oscillated a cylinder. Warui and Fujisawa, [92], devised experimentally a feedback control using the oscillations of the cylinder as an actuator. Sakamoto and Haniu, [81], introduced a smaller cylinder near the main cylinder. Li and Aubry, [59], constructed a feedback controller based on a four-dimensional potential flow model of point vortices to stabilize the lift of a cylinder and control vortex shedding using perturbation and asymptotic methods. Huang, [47], suppressed vortex shedding by the feedback sound. Anderson and co-workers (see e.g, [6]) performed experiments on the vorticity control by oscillating foil. Mahir and Rockwell [62, 63] studied experimentally manipulation of the wake behind two vibrating cylinders. Wang and co-workers, [91], discussed controlling the wake structure by manipulating the separation point thus modifying the Lagrangian coherent structures.

The complexity of wake structure leads to the multiplicity of the optimal control approaches. A typical low order model-based method involves two stages: first the equations of the fluid flow are approximated using reduced order models and then an optimal control algorithm is sought for the reduced order model. The difference among various approaches consists in the choice of the basis functions used to obtain the reduced models and the cost functions that are minimized. In the reduced basis approach one uses as basis functions the terms which arise in series expansion of the solution with respect to a certain parameter (see e.g. Ito and Ravindran, [52]). The proper orthogonal decomposition (POD) approach was applied by Graham et al., [37, 38], and by Afanasiev and Hinze, [4]. Distributed controls were used by Abergel and Temam, [2], Gunzburger and co-workers, [40], Hou and Ravindran, [45, 46]; blowing and suction on the surface of the cylinder was studied by Berggren, [14], Bewley and co-workers, [15], Ghattas and Bark, [33], Li and co-workers, [60]; velocity tracking (boundary velocity controls) was employed by Gunzburger and Manservisi, [42], Gunzburger and co-workers, [41], Hou and Ravindran, [45, 46]. Homescu and co-workers, [44], and Li and co-workers, [60], constructed numerically optimal control based on adjoint method with the actuation being in the form of cylinder rotation and blowing/suction, respectively.

Just as the different methods were used for the actuation of the wake, various flow characteristics were used as the cost functionals. Abergel and Temam, [2], minimized the turbulence for a flow respectively driven by volume forces, a gradient of temperature and a gradient of pressure (the turbulence being measured by a L^2 norm of the vorticity or, respectively, by studying the stress at the boundary); Berggren, [14], minimized the vorticity field. Bewley and co-workers, [15], reduced the turbulent kinetic energy and drag;

Protas, [79], minimized the sum of the work needed to resist the drag force and the work needed to control the flow; Ghattas and Bark, [33], used as an objective function the rate at which energy is dissipated in the fluid. Li and co-workers, [60], used the adjoint approach to obtain the numerical solution to the problem of controlling vortex shedding behind a rotating cylinder for several cost functions. Protas [78] uses the Foppl point vortex system as a reduced-order model for stabilization of the steady symmetric solution in an unstable laminar wake using the LQG (Linear Quadratic Gaussian) approach.

2.4 Vortex Dynamics in Shear and Mixing Layers

In many applications, the main objective is to maximize the transport and mixing of fluid and/or external particles between the different parts of the flow. Typically, the stronger the interaction between the coherent structures of the flow, the more intense the mixing is. However, even if the interaction is strong, the insufficiency (or suboptimality) of the rate of mixing in many applications necessitates the consideration of control algorithms. Here we briefly review research on control of of mixing in mixing layers (e.g. shear layers), where dynamics includes interaction of multiple vortices (in the simplest case, vortex pairs) and the interaction between the recirculation zone (typically consisting of a single vortex) and the outer flow.

It is widely accepted that large quasi-deterministic, vortical, spanwise structures are the essential ‘building blocks’ of the mixing layer and are responsible for virtually all of the momentum transfer across its extent. In a sense, the objective of control of mixing is opposite to other problems of vortex control: instead of stabilization of vortex configurations, the purpose is to destabilize them. Control of the mixing layer started with the work of Hoa and Huang [43], where the spreading rate of the layer was controlled using the subharmonic (relative to the most-amplified frequency) actuation of the merging. In the later papers, the most attention was paid to the selection of frequencies of actuation. Multiple-frequencies approach was suggested by Inoue, [48], who used two (and later in [49] three) distinct frequencies. De Zhou and Wygnanski, [27], studied the multiple-frequencies actuation experimentally. Acharya and co-workers, [3] illustrated that the subharmonic forcing with different frequencies affects the vortex formation and can regularize the shear layer. See [28] and [39] for a review, the latter especially covering experimental methods.

The research on control of mixing between the recirculation zone and the outer flow was initiated with the articles of Noack and co-workers, [72, 73]. The main technique was to modify the motion of the vortex to enhance the exchange between the recirculation zone and the outer flow. The criterium for the efficiency of the mixing was based on the so-called lobe dynamics - that measure the volume of fluid taken into (or out) of the recirculation zone. The optimal vortex motion, was then stabilized by using an observer and a feedback law. An alternative measure – the approach suggested in [26] - is to consider the Kolmogorov-Sinai Entropy. The feedback-control approach for a variety of control of mixing objectives can be found in [1, 12].

2.5 Vortex Control for Improved Swimming Efficiency

In the previous settings we assumed that the vortices were created by the original system itself, either be it away from the observable region, or inside it. However, it is the very *creation* of the vortices that makes certain devices work. Research in locomotion gave birth to a relatively recent application of vorticity control, robotics. Efficiency of fish swimming lead to interest in adapting the fish swimming techniques to swimming robots. Wolfgang and co-workers, [93], numerically illustrated the generation and manipulation of vorticity in a swimming and turning live fish. Zhu and others, [98], studied numerically flow structure and vorticity control in fish-like swimming, in particular vorticity controlled thrust and propulsive efficiency. See also [36, 86, 87] for reviews of recent developments in vorticity control in fish-like swimming.

2.6 Topics Left Out of the Scope

At the end of the current section we would like to mention topics that involve the control of vortical structures near an airfoil, that were left out of the scope of the paper. The reasoning for the omission is that in our view they are not on the immediate line of thought that leads to the control of vortex merging and also the presence of very recent reviews. A throughout reviews of control of airplane trailing vortices, including vortex breakup is presented in [67, 68, 83]. A review of aircraft wake vortices is presented in [32]. And finally a review of control of vorticity aimed at reduction of aircraft drag is contained in [18].

Besides the topics mentioned above, we did not discuss control of turbulence, control of global vortical structures (e.g. Taylor Vortex Flow) and the problems where the changing of vortex structure comes as a byproduct of achieving another goal.

3 Towards Control of Vortex Merging

In the previous section we discussed the domain of applicability of the vortex control methods. We saw that the control of vortex interaction may be a key to manipulating the whole flow structure. Among the processes of vortex interaction, the impact of *vortex merging* is, probably, the largest. In the present section we discuss controlling interaction of two vortices in order to facilitate or prevent merging. It is known (see e.g. [20]) that if the vortex separation is larger than some critical value, vortices do not merge for a long time that depends on the rate of viscous dissipation. On the other hand, if vortices are close they merge fast, usually before they complete a full turn around each other. Therefore, manipulating of the separation is crucial for controlling the merging.

We separate two different cases: when vortices are far apart (in other words, characteristic size of vortex core is much smaller then the characteristic vortex separation) and when vortices are relatively close. In the former case, that we discuss in the next subsection, we use the point vortex approximation. The latter case, when the shape of vorticity distribution is important and vortex merging is possible, we discuss in Sect. 5.

4 Control of a Pair of Point Vortices

In the leading approximation two co-rotating point vortices are in a stable periodic motion. Hence, one can use weak (adiabatic) control, a particular case of singular perturbations control [56]. We review the problem of controlling the state of a pair using a strain field or a field of a single souce/sink. In both cases, when considering the case of weak control we use the method of *averaging* over the fast rotation of vortices around the center of vorticity. For the case of the strain field actuation whose integral over a cycle of vortex rotation is bounded the optimal solutions in integral setting are impulsive – they consist of Dirac delta functions applied at optimal phases (when a system is the most "receptive" to given control inputs) during the cycle of vortex rotation. While these results are obtained by "direct" optimization, For the case of the source-sink field we pose the optimal control problem through the use of the Pontryagin maximum principle on the averaged system. A solution is a set of Dirac delta pulses at optimal phases of the vortex rotation cycle leading to a natural connection between the averaging method and impulsive control [82, 94]. The qualitative aspects of the results presented here are not restricted to the particular form of control flows, but stand for a wide class of actuations. For example, the fact that the source flow profile induced by an actuator in a source/sink experiment might resemble a jet more then a uniform source should not change the conclusion that the optimal control in our specific setting is achieved by a set of pulses synchronized with the internal dynamics of the vortex pair. Detailed analysis, elements of which are shown here, is contained in [90].

4.1 Averaging for a Co-Rotating Vortex Pair

The state of a pair of point vortices positioned at $\mathbf{x}_1, \mathbf{x}_2$ can be described by four variables: the position of the joint vorticity centroid, $\mathbf{x}_c$, the vortex separation, $2r$ and the relative phase of vortex rotation, φ. In terms of $\mathbf{x}_1$ and $\mathbf{x}_2$, the coordinates of $\mathbf{x}_c = (x_c, y_c) = (1/\Gamma)(\Gamma_1\mathbf{x}_1 + \Gamma_2\mathbf{x}_2)$, where Γ_j is the circulation of the j-th vortex and $\Gamma = \Gamma_1 + \Gamma_2$. We restrict our discussion to the case of the same sign vortices (a co-rotating vortex pair). The equations of motion for a co-rotating vortex pair in the absence of any external perturbation are:

$$\begin{aligned}\dot{\mathbf{x}}_c &= 0, \\ \dot{r} &= 0\,, \\ \dot{\varphi} &= \Omega\,(r)\ .\end{aligned} \tag{1}$$

Each vortex moves along a circle of radius r_j centered at $\mathbf{x}_c$ with the same angular frequency $\Omega(r) = \Gamma/8\pi r^2$. $\mathbf{x}_c$ and r are the integrals of motion of the unperturbed system.

Now we assume that an external perturbation (control) in the form of a single-input potential velocity field $\mathbf{u}(\mathbf{x}, t) = \varepsilon\alpha\,(t)\,\mathbf{v}(\mathbf{x})$ of small magnitude $\varepsilon\alpha(t) \ll 1$ is imposed on the vortex pair. The evolution equations for the perturbed system are

$$\begin{aligned}\dot{\mathbf{x}}_c &= \varepsilon\alpha(t)\frac{1}{\Gamma}\sum_{j=1}^{2}\Gamma_j\mathbf{v}_j\ , \\ \dot{r} &= \varepsilon\alpha(t)\sum_{j=1}^{2}\frac{1}{r_j}\left(\mathbf{v}_j, \mathbf{r}_j\right)\ , \\ \dot{\varphi} &= \Omega\,(r) + \varepsilon\alpha(t)\sum_{j=1}^{2}\frac{1}{r_j^2}\left|\mathbf{v}_j \times \mathbf{r}_j\right|\ ,\end{aligned} \tag{2}$$

where $\mathbf{v}_j$ is the value of $\mathbf{v}$ at the position of the j-th vortex. This is a 4-dimensional nonlinear system for which it is hard to obtain direct controllability results. The rate of change of φ is much larger (by a factor of order $1/\varepsilon$) than the rate of change of other variables. Assuming that $\alpha = \alpha(\varphi)$, we can average the perturbation over a fast period and consider the averaged system instead of the exact one. This is in spirit of constructive controllability results using averaging over fast-varying, time-dependence sinusoidal inputs (see e.g. [58, 17]). The difference in our case is that the fast dynamics is provided by the system itself through the angle φ. In addition, the control needs to be specified only for the slow-evolving variables $\mathbf{x}_c$ and r since the desired angle can be reached due to the nominal dynamics of the system (cf. [88]). This is a typical situation for Hamiltonian systems with integrable nominal dynamics [9, 65, 66], where splitting into slow and fast (action and angle) coordinates is possible. The averaged system is

$$\dot{\overline{\mathbf{x}}}_c = \varepsilon\frac{1}{\Gamma}\frac{1}{2\pi}\int_0^{2\pi}\alpha(\varphi)\sum_{j=1}^{2}\Gamma_j\mathbf{v}_j\ d\varphi, \quad \dot{\overline{r}} = \varepsilon\frac{1}{2\pi}\int_0^{2\pi}\alpha(\varphi)\sum_{j=1}^{2}\frac{1}{r_j}\left(\mathbf{v}_j, \mathbf{r}_j\right)\,d\varphi. \tag{3}$$

The solutions of averaged system (3) describe the behavior of $\mathbf{x}_c$ and r of exact system (2) to an accuracy of order ε over time intervals of order $1/\varepsilon$ [10, 16] provided $\Omega(r)$ is not zero, in the case of sufficiently smooth right-hand sides.

In the following sections we discuss solution to problems of controlling the position of a vortex pair using simple potential velocity fields $\mathbf{v}$: strain field and source/sink field. Our analysis relies heavily on averaged equations (3) and is prompted but by smallness of the control input magnitude $\varepsilon\alpha$ with respect to angular frequency of rotation Ω.

4.2 Two Vortices Controlled by a Strain Field

The first problem we discuss is controlling the state of a vortex pair using a single strain field described by a potential $\Psi = \varepsilon\alpha\left(\varphi(t)\right)xy$ and the corresponding velocity field $\mathbf{u}$:

$$u_x = \varepsilon\alpha\left(\varphi(t)\right)x, \quad u_y = -\varepsilon\alpha\left(\varphi(t)\right)y. \tag{4}$$

It will prove useful to represent the control α as

$$\alpha = \langle\alpha\rangle + \alpha_f\left(\varphi\right). \tag{5}$$

where $\langle\alpha\rangle = (\int_0^{2\pi}\alpha\, d\varphi)/2\pi$. Substituting (4) into (2) and averaging over a fast period we arrive at:

$$\begin{aligned}
\dot{\overline{x}}_c &= \varepsilon\overline{x}_c\frac{1}{2\pi}\int_0^{2\pi}\alpha\, d\varphi = \varepsilon\overline{x}_c\langle\alpha\rangle, \\
\dot{\overline{y}}_c &= -\varepsilon\overline{y}_c\,\frac{1}{2\pi}\int_0^{2\pi}\alpha\, d\varphi = -\varepsilon\overline{y}_c\langle\alpha\rangle, \\
\dot{\overline{r}} &= \varepsilon\overline{r}\frac{1}{2\pi}\int_0^{2\pi}\alpha\cos(2\varphi)\, d\varphi = \varepsilon\overline{r}\frac{1}{2\pi}\int_0^{2\pi}\alpha_f(\varphi)\cos(2\varphi)\, d\varphi.
\end{aligned} \tag{6}$$

Note, that the evolution equations for $\overline{\mathbf{x}}_c$ and $\overline{r}$ are separated in a sense that $\dot{\overline{\mathbf{x}}}_c$ depends on the average part of α only, while $\dot{\overline{r}}$ depends on the oscillatory part of α. The separability is not a property peculiar to strain field controlled dynamics of a vortex paper. The same property holds in a generic N actions + 1 angle system.

We now want to study controllability of (6): is it possible to change both $\overline{\mathbf{x}}_c$ and $\overline{r}$ from some arbitrary initial to arbitrary final position by using some $\alpha(\varphi)$? We divide the algorithm into two steps: (1) Moving $\overline{\mathbf{x}}_c$ from the initial to the final position without changing r much; (2) Adjusting r without disturbing $\overline{\mathbf{x}}_c$ much. The combination of the two steps specifies the required control. Note, that the two steps can be combine into one as the evolution of $\overline{\mathbf{x}}_c$ and $\overline{r}$ depend on different terms in expansion (5).

Step 1. A strain field is specified if we prescribe the position of the origin, the orientation and the strength. It is clear that if both the origin and the orientation of are prescribed, $\overline{\mathbf{x}}_c$ can not be moved at will across the plane. However, given the initial and final state we can prescribe either the initial orientation or position of the origin of the strain field such that the system is moved from the initial to the final state in finite time. Varying $\langle\alpha\rangle$ we just change the speed of $\overline{\mathbf{x}}_c$ without changing the trajectory. For $\langle\alpha\rangle = \text{const}$, it takes the time $\tau \propto 1/(\varepsilon\,\langle\alpha\rangle)$ to move $\overline{\mathbf{x}}_c$ from the initial to the final position.

It follows from (6), that if $\alpha_f(\varphi) = 0$ in the averaged system the vortex separation does not change. As the duration of Step 1, τ, is of order $1/\varepsilon$, the vortex separation in exact system is within ε from the one in the averaged system, and, therefore, from the initial value. Thus the objective of Step 1 is met. For numerical simulations that check the numerical value of error of the averaging technique see [89].

Step 2. Now our aim is to change the vortex separation from an initial value $2\overline{r}_i$ to a final value $2\overline{r}_f$ without moving $\overline{\mathbf{x}}_c$. We can do it by keeping the perturbation in resonance with the unperturbed system: it follows from the last line in (6), that if $\alpha_f(\varphi)$ is of the same (opposite) sign as $\cos(2\varphi)$, $\dot{\overline{r}}$ is always positive (negative) for non-vanishing values of $\overline{r}$. It follows from the first two lines in (6), that if we set $\langle\alpha\rangle = 0$ for this step $\overline{\mathbf{x}}_c$ in the averaged system stays the same. The duration of Step 2, τ, is of order $1/\varepsilon$,r and $\mathbf{x}_c$ in exact system are within ε from the ones in the averaged system.

Step 2 provides a convenient setting to describe the optimal control for the averaged settings. Define a cost function as a flow of a control field through some surface. The flow is proportional to

$$\Phi(\tau) = \varepsilon \int_0^\tau |\alpha|\; dt\,, \tag{7}$$

where the integration is performed over the entire evolution of the system ($\tau \sim 1/\varepsilon$). It is clear from (6), that as during Step 2 we do not want to move the center of vorticity, without loss of generality, we can set $\langle\alpha\rangle = 0,\ \alpha = \alpha_f(\varphi)$. It follows from (7), that $\dot{\Phi}(t) = \varepsilon\,|\alpha|\,(t)$. Averaging this relation over φ we get

$$\dot{\overline{\Phi}} = \varepsilon \frac{1}{2\pi} \int_0^{2\pi} |\alpha(\varphi)|\; d\varphi.$$

Introducing $\overline{\Phi}$ as a new (slow) time we can rewrite the last line in (6) as

$$\frac{d\overline{r}}{d\overline{\Phi}} = \overline{r}\, \frac{\int_0^{2\pi} \alpha_f(\varphi)\cos(2\varphi)\; d\varphi}{\int_0^{2\pi} |\alpha_f(\varphi)|\; d\varphi} \tag{8}$$

under the constraint $\langle\alpha_f\rangle = 0$. We reduced the problem of minimizing flow to minimizing the time of the evolution. It follows from (8), that the final value of Φ does not depend on ε. Therefore, one can not just reduce ε to improve the efficiency of control. On the other hand, the value of τ (see (7)) is (in the first approximation) inversely proportional to ε. All we have to determine is the shape of $\alpha_f(\varphi)$. It follows from (8), to minimize the total flux we need to maximize th integral at the right-hand-side over all α. It was shown in [89] that such a function $\alpha_f(\varphi)$ is a set of pulses which coincide with the extrema of $\cos 2\varphi$. For example, $\alpha_f(\varphi)$ can be chosen as

$$\alpha_f(\varphi) = \pm\frac{1}{4}\left(\delta(\varphi) + \delta(\varphi - \pi) - \delta(\varphi - \pi/2) - \delta(\varphi - 3\pi/2)\right) , \qquad (9)$$

where the positive and negative signs correspond to $\overline{r}_f > \overline{r}_i$ and $\overline{r}_f < \overline{r}_i$, respectively. The control given by equation (9) has the form of the so-called impulsive control [94, 82]. Thus we have obtained a natural connection between optimal control in the averaging context and impulsive control methods. Note that the resulting control (*control of vortex merger by minimum flux*) is quite different from the standard bang-bang control profile (that would be appropriate in the context of the *minimum time* problem, see e.g. [54]). The detailed discussion of this is contained in [90].

Strictly speaking, the resulting optimal control $\overline{\alpha}_f(\varphi)$ is not in the space of functions but in the space of signed measures (Dirac δ-functions) on the circle and it does not satisfy the requirements used in [10, 16] to prove the validity of the averaging method. However, the smoothness requirement in φ that these references use can be easily overcome for the impulsive control (9) using integral methods of averaging theorem proof such as in [61] to prove that the full system and the averaged system stay within ε over times of order $1/\varepsilon$.

4.3 Two Vortices in a Source/Sink Field

In this section we consider the problem of controlling the separation of a vortex pair using a single source/sink field. Whilst in the previous section optimality of a particular control was shown using "direct" optimization methods, here the problem becomes more complicated and requires the use of more sophisticated tools from control theory, such as the Pontryagin maximum principle. The control is still specified as a feedback, but in this case besides the angle φ the control depends on both the vortex separation and the position of the center of vorticity. We are interested only in the vortex separation problem, – the final position of the center of vorticity $\mathbf{x}_c$ is arbitrary.

Let the source/sink be at the origin. Then the control flow is described by a potential

$$\Psi = \varepsilon\alpha\left(\mathbf{x}_c(t), r(t), \varphi(t)\right) \arctan\left(y/x\right) .$$

Like in the previous section we consider the control fields that do not depend explicitly on time. Positive and negative values of α correspond to a source and a sink, respectively. The corresponding velocity field is

$$u_x = \varepsilon\alpha\left(\mathbf{x}_c, r, \varphi\right)\frac{x}{x^2 + y^2}, \qquad u_y = \varepsilon\alpha\left(\mathbf{x}_c, r, \varphi\right)\frac{y}{x^2 + y^2} .$$

Let the instantaneous position of the center of vorticity, $\mathbf{x}_c$, be at $x = c$, $y = 0$. Denote by φ the phase of rotation of the vortices: $\tan\varphi = (x_1 - c)/y_1$. We also assume that the source/sink is removed from the vortices: $c > r$ throughout the evolution. For the sake of simplicity we consider only symmetric in φ perturbations that do not move $\mathbf{x}_c$ from the x-axis and also reduce the discussion to a particular case of two identical vortices. Substituting the above control field into (2) and averaging over the fast period we get

$$\dot{r} = \varepsilon \frac{1}{2\pi} \int_0^{2\pi} \alpha\left(\mathbf{x}_c, r, \varphi\right) f_r \, d\varphi, \quad \dot{c} = \varepsilon \frac{1}{2\pi} \int_0^{2\pi} \alpha\left(\mathbf{x}_c, r, \varphi\right) f_c \, d\varphi, \tag{10}$$

where

$$f_r = r \frac{r^2 - c^2 \cos(2\varphi)}{r^4 + c^4 - 2r^2c^2 \cos(2\varphi)}, \quad f_c = c \frac{c^2 - r^2 \cos(2\varphi)}{r^4 + c^4 - 2r^2c^2 \cos(2\varphi)}.$$

One can see that f_r and f_c satisfy the following relation:

$$cf_c + rf_r = 1. \tag{11}$$

Similarly to Sect. 4.2, c and r can be controlled independently. In what follows, we are going to concentrate on controlling r only. As f_r is periodic in φ, by appropriate choice of $\alpha(\varphi(t))$ we can assure that in the averaged system $\dot{r}$ is always positive (negative) and it vanishes for $r = 0$ only. Hence, we can change r in the averaged system from any initial value to any final value over the time of order $1/\varepsilon$. And, as the exact system stays within ε from the exact system, it means that we can make the vortex separation in the exact system arbitrarily close to the target value.

Similarly to Sect. 4.2, the extrema of f_r and f_c, that coincide and are located at $\varphi = 0$, π and $\pm\pi/2$, determine the structure of optimal control. Define the optimal control as a perturbation that minimizes the total flow through the source/sink and satisfies all the imposed constrains. The flow is given by

$$\Phi = \varepsilon \int_0^{\tau} |\alpha| \, dt,$$

where the integration is performed over the entire evolution of the system ($\tau \sim 1/\varepsilon$). We consider the problem of changing the vortex separation from an initial value $2r_i$ at $t = 0$ to a final value $2r_f$.

Evolution equations (10) are nonlinear and possess 2 degrees of freedom. Therefore, the solution is more complicated compared with the strain field case. What allows us to obtain an analytical solution is conservation law (10). Similar to the strain field case, the optimal control is a set of δ-pulses. If $r_f > r_i$, the optimal control consists of sink pulses applied at $\varphi = 0, \pi$. For $r_f < r_i$ there exists a critical value of r, $r_* = r_f \, (1 + r_i^2/c_i^2)^{1/2}$, such that depending on the relation between r_i and r_* we can have no switches or one switch. For $r_i < r_*$ there are no switches and the control is a set of source pulses at $\varphi = 0, \pi$. For $r_i > r_*$ the control is a set of sink pulses at $\varphi = \pm\pi/2$ until r reaches the value $r = r_*$ and then the control becomes a set of source pulses at $\varphi = 0, \pi$.

5 Control of a Pair of Vortex Patches

When the vortices come closer the point vortex approximation cease to be valid and one must take into account distributed vorticity. A possible approximation is vortex patches (blobs) model. The control of such structures was pioneered by Cortelezzi and his co-workers [25]. In the current section we study the behavior of identical patches. We provide insight into using control to force or prevent the merging, by putting a point vortex of a time-varying strength in the joint center of the vorticity.

For a certain range of parameters (e.g. initial separation) merging might occur or not occur depending on the initial conditions. In other words, in the phase portrait there are regions of merging and co-rotating regimes. Depending on the initial and current location of a phase point different control regimes are admissible. While in a co-rotating state the time of evolution might be of no concern, one need to act pretty fast to prevent patches on a merging trajectory from passing beyond a point of no return (although a recent publication by Yu and Driscoll, [96], suggest that even after the beginning of merging the process can be somewhat reversed).

Consider uncontrolled nominal dynamics of a pair of two identical elliptical vortex patches. Numerous simulations showed, that, if the separation between the patches is relatively large, then initially elliptical patches stay close to (albeit not exactly in) the elliptic shape. Therefore, we can use the characteristics of the ellipses to describe, with varying degree of accuracy, the system. Any state of such system is specified by three variables: the separation between the centers of the patches, $2R$, the aspect ratio $\lambda = b/a$, where a and b are the lengths of the major and minor axes, respectively, and the orientation angle φ (see Fig. 1). The system possesses two conserved quantities. The first is the position of the joint center of vorticity: $\mathbf{x_c} = (\mathbf{x_1} + \mathbf{x_2})/2$, where $\mathbf{x_i}$ is the position of the center of the i-th patch, assumed to be located at the origin, the way it is shown in Fig. 1. The second is the total angular impulse, $M = 2A\omega\left(R^2 + (A/4\pi)\left(1 + \lambda^2\right)/\lambda\right)$, where A and ω are the area and the circulation of a patch, respectively. Note, that the angular momentum (in a more general, integral, form) is conserved even without the ellipticity assumption. The value of M roughly specifies how far (compared with their size) from each other the patches are located. It follows from the conservation of M, that there is a maximum value of the patches' separation, which is achieved when $\lambda = 1$ (circular patches) and is given by $R^2_{\mathrm{max}} = M/(2A\omega) - A/(2\pi)$.

We discuss two possible control algorithms. The first is obtained using method of *flat coordinates* [11, 71]. We show that Hamiltonian structure of the *control field* is important to prove controllability. The Hamiltonian structure of the *nominal system* leads to significant extension of reachable domain. In particular, it provides a way to overcome singularities. The second method is to use the method of adiabatic control, similar to the one discussed in the previous subsection. Using this problem as an example we compare these approaches and discuss there relative advantages and disadvantages.

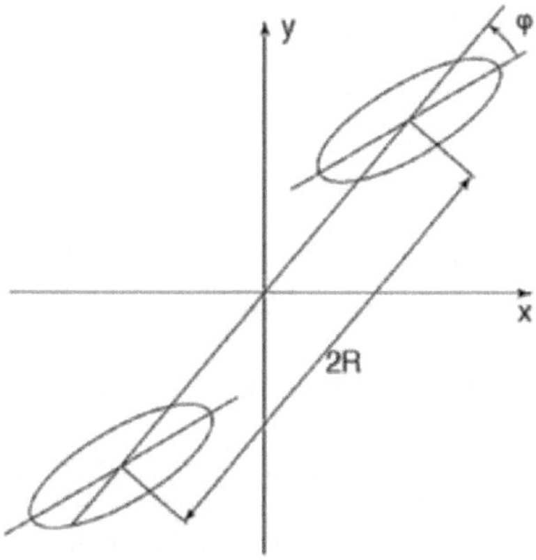

Fig. 1. A sketch of two elliptical vortex patches symmetrically situated around the joint center of vorticity at the origin

In the normalized variables evolution equations can be written as

$$\dot{\lambda} = +\frac{1}{4\rho}\lambda \sin 2\varphi + \gamma(t)\frac{1}{\rho}\lambda \sin 2\varphi, \tag{12}$$

$$\begin{aligned}\dot{\varphi} &= \frac{\lambda}{(1+\lambda)^2} - \frac{1}{8\rho}\left[2 - \left[\frac{1+\lambda^2}{1-\lambda^2} + \frac{1}{4\rho}\frac{1-\lambda^2}{\lambda}\right]\cos 2\varphi\right] \\ &\quad -\gamma(t)\frac{1}{\rho}\left(1 - \frac{1+\lambda^2}{1-\lambda^2}\cos 2\varphi\right),\end{aligned} \tag{13}$$

$$4\rho = \sigma - \left(\lambda + \frac{1}{\lambda}\right).$$

where $\rho = R^2\pi/A$ and $\sigma = 2\pi M/A^2\omega$ are normalized square of the patches separation and angular impulse, respectively, λ is the aspect ratio and φ is the orientation angle (see Fig. 1). Normalized vorticity of a point vortex, $\gamma(t)$, is a control parameter.

Terms without γ describe nominal dynamics. Of them, the first term on the right-hand side in the second line is self-interaction and the remaining terms are mutual interaction between the patches. The dynamics of the nominal system was studied in details in [64, 95].

5.1 Flat Coordinates

When the patches are in the merging-bound state, the major requirement for a control algorithm is to be fast. It also might be quite strong control in order to overcome dynamics of the original system and make the state variables evolve in a prescribed, however "unnatural", manner. In that case, of course, the structure of the nominal flow is completely destroyed. The method of *flat coordinates* [11, 71] provides a way to construct such an algorithm. Apply change of variables

$$z_1 = -\frac{1}{2}\frac{1-\lambda^2}{\lambda}\cos 2\varphi + \frac{(1-\lambda)^2}{\lambda},$$
$$z_2 = \sin 2\varphi\left(\frac{1-\lambda}{1+\lambda} - \frac{1}{32\rho^2}\left(\frac{1-\lambda^2}{\lambda}\right)^2\cos 2\varphi\right). \tag{14}$$

In the new coordinates evolution equations (12) and (13) can be written as

$$\dot{z}_1 = z_2, \quad \dot{z}_2 = g_1(z_1, z_2) + \gamma(t) g_2(z_1, z_2), \tag{15}$$

where $g_1(z_1, z_2)$ and $g_2(z_1, z_2)$ are certain (known) functions of z_1 and z_2. Advantage of new variables (14) is that in (15) $\gamma(t)$ appears in one equation only. It is clear that the required change of variables exist if and only if the control field is integrable. If the control field has just one degree of freedom, it is integrable if it is Hamiltonian. The method of flat coordinates works as follows. The initial and target states of the system specify the initial and terminal values of z_1 and z_2. We can choose an arbitrary trajectory $z_1(t)$ provided that it satisfies the first equation in (15) at the initial and final points. The evolution of z_2 is governed by the first equation in (15). The second equation in (15) yields the equation for the control parameter $\gamma(t)$:

$$\gamma(t) = \frac{\dot{z}_2 - g_1(z_1, z_2)}{g_2(z_1, z_2)},$$

provided that $g_2(z_1, z_2)$ is not zero. The difference between the system under consideration and those studied earlier (see e.g. [72]) is that $g_2(z_1, z_2)$ is zero on a special curve, which we call the *uncertainty curve* and denote by Γ. The uncertainty curve presents an obstacle for the method of flat coordinates: the equation for $\gamma(t)$ has a singularity on that curve. Therefore, we need to modify the method of flat coordinates.

If the initial and final points are on the same side of the uncertainty curve we can choose a path $z_1(t), z_2(t)$ that does not intersect the uncertainty curve. If these points reside on different sides of the uncertainty curve, we can navigate the way across Γ using the internal dynamics of the original system. First, we bring the system close to the uncertainty curve in a region where the vector field of the original system intersects the uncertainty curve in the proper direction: inside or outside. The existence of such a region follows from the preservation of the phase volume in Hamiltonian systems. Second, we turn the control off ($\gamma = 0$). The original flow moves the system across the uncertainty curve. Then we proceed in a conventional way with flat coordinates using the terminal point of the previous step as the new initial point.

To summarize, the usefulness of the method of flat coordinates for flow control lies in the freedom to choose nearly arbitrary control profile. In particular, we can avoid "dangerous" parts of the phase space and make the total time of the evolution arbitrary small. These properties are particularly important if the objective is to take the system from a merging-bound to a co-rotating state. In that case the available time is obviously bounded: patches will merge in less then half a period of full rotation. Therefore, although the method of flat coordinates is not cost-efficient, it is quite useful as "emergency" control.

5.2 Adiabatic Control

The control mechanism considered in Subsect. 5.1 could involve relatively big control impacts (large values of $\gamma(t)$) as they are supposed to suppress the nominal dynamics. In the present section we discuss the opposite case, when control can be considered to be small perturbation compared with the original system. In this case characteristic time scale of the unperturbed system is much smaller than characteristic time scale of the perturbation. This means that we can apply the method only as long as patches are in some form of steady state, i.e. in a non-merging regime [64]. This approximation allows us to average the perturbation over a fast period and consider an averaged system instead of the exact one.

The state of the unperturbed system can be described using H_0, which is the excess energy and the Hamiltonian of the nominal system, and an angle variable ψ, that is a phase on a line of constant H_0. The equations of motion of nominal dynamics are:

$$\dot{H}_0 = 0, \quad \dot{\psi} = \Omega\,(H_0)\,, \tag{16}$$

where $\Omega\,(H_0)$ is angular frequency. Note, that Ω is a function of H_0 only.

Let the control vortex have a characteristical strength $\varepsilon \ll 1$. The evolution equations for the perturbed system are

$$\dot{H}_0 = \varepsilon\gamma(t) f_H, \quad \dot{\psi} = \Omega\,(H_0) + \mathcal{O}(\varepsilon), \tag{17}$$

where f_H is the time derivative of H_0 due to the control field:

$$f_H = -\sin 2\varphi \left(\frac{1-\lambda}{1+\lambda} - \frac{1}{32\rho^2} \left(\frac{1-\lambda^2}{\lambda} \right)^2 \cos 2\varphi \right).$$

Averaging (17) over a fast period we arrive at

$$\dot{H}_0 = \varepsilon \frac{1}{2\pi} \int_0^{2\pi} \gamma(\psi) f_H(\psi)\, d\psi. \tag{18}$$

The solutions of averaged system (18) are close to those of the nominal system (17) to an accuracy of order ε over time intervals of order $1/\varepsilon$ [10, 16] provided $\Omega(H_0)$ is not zero.

As f_H is π-periodic in ψ, it follows from (18), that we can always choose $\gamma(\psi)$ such that H_0 will change from any initial value to any final value, provided that the system stays in a steady, non-merging regime. Therefore, we can use the feedback control and keep γ to be of the same (opposite) sign as f_H to make H_0 increase (decrease). This method allows us to move the system to the desired phase curve of the original system and the fast rotation will do the rest.

Similar to the previous section, consider a cost function of the form similar to (7):

$$\Phi = \varepsilon \int_0^T |\gamma(t)|\, dt, \tag{19}$$

Note that $\dot{\Phi}(t) = \varepsilon\, |\gamma|\,(t)$. By averaging this relation over ψ and introducing Φ as a new (slow) time we arrive at the equivalent of (8):

$$\frac{dH_0}{d\Phi} = \frac{\int_0^{2\pi} \gamma(\psi) f_H(\psi)\, d\psi}{\int_0^{2\pi} |\gamma(\psi)|\, d\psi}. \tag{20}$$

Similarly to (9), the optimal control can be written as a set of Dirac δ-functions:

$$\gamma_*(\psi) = \pm \sum_i \operatorname{sign}(f_H(\psi))\, \delta\left(\psi - \psi_{\max}^{(i)}\right), \tag{21}$$

where $\{\psi_{\max}^{(i)}(H_0)\}$ is a set of values of ψ at which $|f_H|$ reaches its maximum value, $f_{H,*}$ at a given value of H_0; plus and minus signs correspond to increase and decrease of H_0, respectively. Substituting (21) into (20) we arrive at

$$\Phi = \int_{H_{0,\min}}^{H_{0,\max}} \frac{1}{f_{H,*}(H_0)}\, dH_0. \tag{22}$$

Optimal control lets the nominal system go around the phase trajectory and the impulses are applied every time the phase point intersects comes to $\psi = \psi_{\max}(H_0)$ to move the system to the next level of H_0.

Note that there is no lower bound of the magnitude of the actuation: when ε becomes smaller, the time of the evolution increases, but the controllability results do not change.

Conclusions

In the present paper we presented an overview of the control of vortex-dominated flows and discussed two particular algorithms of controlling the motion of a pair of interacting vortical structures: point vortices and elliptical vortex patches with application to control of the merging. One way is to use the method of flat coordinates. The Hamiltonian structure of control field is crucial to prove controllability and the Hamiltonian structure of nominal system leads to significant extension of the reachable domain. The other approach is to use the "adiabatic" control, that is a small perturbation compared with the nominal system. In this context, we use the method of *averaging* over the fast rotation of vortices around the center of vorticity. We show that one can make control more efficient by using the internal dynamics of the system, in particular, keeping the perturbation in phase with fast (angle) variable of the nominal system.

References

[1] O.M. Aamo, M. Krstić, and T.R. Bewley. Control of mixing by boundary feedback in 2d channel flow. *Automatica*, 39:1597–1606, 2003.

[2] F. Abergel and R. Temam. On some control problems in fluid mechanics. *Theoretical and Computational Problems in Fluid Dynamics*, 1:303–325, 1990.

[3] S. Acharya, T.A. Myrum, and S. Inamdar. Subharmonic excitation of the shear-layer between 2 ribs – vortex interaction and pressure field. *AIAA Journal*, 29:1390–1399, 1991.

[4] K. Afanasiev and M. Hinze. Adaptive control of a wake flow using Proper Orthogonal Decomposition. Technische Universitat Berlin Preprint, 1999.

[5] C.R. Anderson, Y.C. Chen, and J.S. Gibson. Control and identification of vortex wakes. *Journal of Dynamic Systems Measurement and Control-Transactions of the ASME*, 122:298–305, 2000.

[6] J.M. Anderson, K. Streitlien, D.S. Barrett, and M.S. Triantafyllou. Oscillating foils of high propulsive efficiency. *Journal of Fluid Mechanics*, 360:41–72, 1998.

[7] I. Aranson, H. Levine, and L. Tsimring. Controlling spatiotemporal chaos. *Physical Review Lett.*, 72:2561–2564, 1994.

[8] H. Aref. Integrable, chaotic, and turbulent vortex motion in two-dimensional flows. *Annual Review of Fluid Mechanics*, 15:345–389, 1983.

[9] V.I. Arnold. *Mathematical Methods of Classical Mechanics.* Springer-Verlag, New York, 1978.

[10] V.I. Arnold. *Geometrical Methods in the Theory of Ordinary Differential Equations*, volume XI. New York-Heidelberg-Berlin: Springer-Verlag, New York-Heidelberg-Berlin, 1983.

[11] J. Baillieul. The geometry of controlled mechanical systems. *In Mathematical Control Theory, J. Baillieul and J.C. Williams, Eds.*, pages 322–354, 1999.

[12] B. Bamieh and I. Mezić. A framework for destabilization of dynamical systems and mixing enhancement. *Proc. 30th IEEE CDC*, page Paper 4980, 2001.

[13] G. K. Batchelor. *An introduction to Fluid Dynamics.* Cambridge University Press, Cambridge, 1967.

[14] P. Berggren. Numerical solution of a flow control problem: vorticity reduction by dynamic boundary action. *SIAM Journal in Control and Optimization*, 19:829–860, 1998.

[15] T.R. Bewley, R. Temam, and M. Ziane. A general framework for robust control in fluid mechanics. *Physica D*, 138:360–392, 2000.

[16] N.N. Bogolyubov and Yu.A. Mitropolsky. *Asymptotic Methods in the Theory of Nonlinear Oscillations*, volume 537. Gordon and Breach Science Publ., New York, 1961.

[17] F. Bullo. Averaging and vibrational control of mechanical systems. *SIAM Journal on Control and Optimization*, 41:542–562, 1999.

[18] D.M. Bushnell. Aircraft drag reduction – a review. *Proceedings of The institution of Mechanical Engineers Part G – Journal of Aerospace Engineering*, 217:1–18, 2003.
[19] X. Carton, G. Maze, and B. Legras. A two-dimensional vortex merger in an external strain field. *J. of Turbulence*, 3, art. no.045, 2002.
[20] C. Cerretelli and C.H.K. Williamson. The physical mechanism for vortex merging. *Journal of Fluid Mechanics*, 475:41–77, 2003.
[21] S.I. Chernyshenko. Stabilization of trapped vortices by alternating blowing suction. *Physics of Fluids*, 7:802–807, 1995.
[22] S.S. Collis, K. Ghayour, M. Heinkenschloss, M. Ulbrich, and S. Ulbrich. Optimal control of unsteady compressible viscous flows. *International Journal for Numerical Methods in Fluids*, 40:1401–1429, 2002.
[23] L. Cortelezzi. Nonlinear feedback control of the wake past a plate with a suction point on the downstream wall. *Journal of Fluid Mechanics*, 327:303–324, 1996.
[24] L. Cortelezzi, Y.C. Chen, and H.L. Chang. Nonlinear feedback control of the wake past a plate: From a low-order model to a higher-order model. *Physics of Fluids*, 9:2009–2022, 1997.
[25] L. Cortelezzi, A. Leonard, and J.C. Doyle. An example of active circulation control of the unsteady separated flow past a semiinfinite plate. *Journal of Fluid Mechanics*, 260:127–154, 1994.
[26] D. D'Alessandro, M. Dahleh, and I. Mezić. Control of mixing in fluid flow: A maximun entropy approach. *IEEE Transactions on Automatic Control*, 44:1852–1863, 1999.
[27] Ming de Zhou and I. Wygnanski. The response of a mixing layer formed between parallel streams to a concomitant excitation at two frequencies. *Journal of Fluid Mechanics*, 441:139–168, 2001.
[28] M. Gad el Hak and D.M. Bushnell. Separation control: review. *J. of Fluid Engineering*, 113:5–30, 1991.
[29] L. Friedland. Control of kirchhoff vortices by a resonant strain. *Physical Review E*, 59:4106–4111, 1999.
[30] L. Friedland and A.G. Shagalov. Resonant formation and control of 2d symmetric vortex waves. *Physical Review Lett.*, 85:2941–2944, 2000.
[31] L. Friedland and A.G. Shagalov. Emergence of nonuniform v-states by synchronization. *Physics of Fluids*, 14:3074–3086, 2002.
[32] T. Gerz, F. Holzapfel, and D. Darracq. Commercial aircraft wake vortices. *Progress in Aerospace Sciences*, 38:181–208, 2002.
[33] O. Ghattas and J. Bark. Optimal control of 2-d and 3-d incompressible navier-stokes flows. *Journal of Computational Physics*, 136:231–244, 1997.
[34] E.A. Gillies. Low-dimensional control of the circular cylinder wake. *Journal of Fluid Mechanics*, 371:157–178, 1998.
[35] E.A. Gillies. Multiple sensor control of vortex shedding. *AIAA Journal*, 39:748–750, 2001.
[36] R. Gopalkrishnan, M.S. Triantafyllou, G.S. Triantafyllou, and D. Barrett. Active vorticity control in a shear-flow using a flapping foil. *Journal of Fluid Mechanics*, 274:1–21, 1994.
[37] W.R. Graham, J. Peraire, and K.Y. Tang. Optimal control of vortex shedding using low-order models. part i – open-loop model development. *International Journal for Numerical Methods in Engineering*, 44:945–972, 1999.

[38] W.R. Graham, J. Peraire, and K.Y. Tang. Optimal control of vortex shedding using low-order models. part ii – model-based control. *International Journal for Numerical Methods in Engineering*, 44:973–990, 1999.

[39] D. Greenblatt and I.J. Wygnanski. The control of flow separation by periodic excitation. *Progress in Aerospace Sciences*, 36:487–545, 2000.

[40] M. Gunzburger, L. Hou, and T. Svobodny T. Analysis and finite element approximation of optimal control problems for the stationary Navier-Stokes equations with distributed and neumann controls. *Mathematics of Computation*, 57:123–151, 1991.

[41] M. Gunzburger, L. Hou, and T. Svobodny T. Boundary velocity control of incompressible flow with an application to viscous drag reduction. *SIAM Journal of Control and Optimization*, 30:167–181, 1992.

[42] M. Gunzburger and S. Manservisi. The velocity tracking problem for Navier-Stokes flows with bounded distributed controls. *SIAM Journal of Control and Optimization*, 37:1913–1945, 1999.

[43] C.M. Ho and L.S. Huang. Subharmonics and vortex merging in mixing layers. *Journal of Fluid Mechanics*, 38:43–69, 1982.

[44] C. Homescu, I.M. Navon, and Z. Li. Suppression of vortex shedding for flow around a circular cylinder using optimal control. *International Journal For Numerical Methods in Fluids*, 119:443–473, 2002.

[45] L. Hou, S.S. Ravindran, and Y. Yan. Numerical solutions of optimal distributed control problems for incompressible flows. *International Journal of Computational Fluid Dynamics*, 8:99–114, 1997.

[46] L. Hou and Y. Yan. Dynamics and approximations of a velocity tracking problem for the Navier-Stokes flows with piecewise distributed controls. *SIAM Journal on Control and Optimization*, 35:1847–1885, 1997.

[47] X. Huang. Feedback control of vortex shedding from a circular cylinder. *Experiments in Fluids*, 20:218–224, 1996.

[48] O. Inoue. Double-frequency forcing on spatially growing mixing layers. *Journal of Fluid Mechanics*, 234:553–581, 1992.

[49] O. Inoue. Note on multiple-frequency forcing on mixing layers. *Fluid Dynamics Research*, 16:161–172, 1995.

[50] A. Iollo and L. Zannetti. Optimal control of a vortex trapped by an airfoil with a cavity. *Flow, Turbulence & Combustion*, 65:417–30, 2000.

[51] A. Iollo and L. Zannetti. Trapped vortex optimal control by suction and blowing at the wall. *European Journal of Mechanics B-Fluids*, 20:7–24, 2001.

[52] K. Ito and S.S. Ravindran. A reduced-order method for simulation and control of fluid flows. *Journal of Computational Physics*, 143:403–425, 1998.

[53] J.B. Kadtke, A. Pentek, and G. Pedrizzetti. Controlled capture of a continuous vorticity distribution. *Physics Letters A*, 204:108–114, 1995.

[54] D. E. Kirk. *Optimal Control Theory: An Introduction.* Prentice Hall, New York, 1970.

[55] K. Koenig and A. Roshko. An experimental study of geometrical effects on the drag and flow field of two bluff bodies separated by a gap. *Journal of Fluid Mechanics*, 156:167, 1985.

[56] P.V. Kokotovic, H.K. Khalil, and J. O'Reilly. *Singular perturbation methods in control: analysis and design.* Academic, London, 1986.

[57] K. Kwon and H. Choi. Control of laminar vortex shedding behind a circular cylinder using splitter plates. *Physics of Fluids*, 8:479–486, 1996.

[58] N. E. Leonard and P. S. Krishnaprasad. Motion control of drift-free left-invariant systems on Lie groups. *IEEE Transactions on Automatic Control*, 40:1539–1554, 1995.
[59] F. Li and N. Aubry. Feedback control of a flow past a cylinder via transverse motion. *Physics of Fluids*, 15:2163–2176, 2003.
[60] Z.J. Li, I.M. Navon, M.Y. Hussaini, and F.A. Le Dimet. Optimal control of cylinder wakes via suction and blowing. *Computers & Fluids*, 32:149–171, 2003.
[61] P. Lochak, C. Meunier, and H.S. Dumas. *Multiphase averaging for classical systems: with applications to adiabatic theorems*. Springer-Verlag, New York, 1988.
[62] N. Mahir and D. Rockwell. Vortex formation from a forced system of two cylinders. part i: tandem arrangement. *Journal of Fluids and Structures*, 10:473–489, 1996.
[63] N. Mahir and D. Rockwell. Vortex formation from a forced system of two cylinders. part ii: side-by-side arrangement. *Journal of Fluids and Structures*, 10:491–500, 1996.
[64] M.V. Melander, N.J. Zabusky, and J.C. McWilliams. Symmetric vortex merger in two dimensions: causes and conditions. *Journal of Fluid Mechanics*, 195:303–340, 1988.
[65] I. Mezić. Controllability, integrability and ergodicity. In *Proceedings of the Mohammed Dahleh Symposium, in Lecture Notes in Control and Information Sciences*, volume 289, pages 213–229. Springer-Verlag, 2003.
[66] I. Mezić. Controllability of hamiltonian systems with drift: Action-angle variables and ergodic partition. In *Proceedings of Conference on Decision and Control, Maui*, 2003.
[67] A.M. Mitchell, D. Barberis, P. Molton, and J. Delery. Control of leading-edge vortex breakdown by trailing edge injection. *Journal of Aircraft*, 39:221–226, 2002.
[68] A.M. Mitchell and J. Delery. Research into vortex breakdown control. *Progress in Aerospace Sciences*, 37:385–418, 2002.
[69] V.J. Modi. Moving surface boundary-layer control: a review. *Journal of Fluids and Structures*, 10:491–500, 1996.
[70] P. K. Newton. *Vortex Dynamics*. Springer-Verlag, New York, 2002.
[71] H. Nijmeijer and A.J. van der Schaft. *Nonlinear dynamical control systems*. Springer-Verlag, New York, 1990.
[72] B.R. Noack and I. Mezićand A. Banaszuk. Controling vortex motion and chaotic advection. In *Proceedings of the 39th IEEE Conference on Decision and Control*, pages 1716–1723, 2000.
[73] B.R. Noack, I. Mezić, G. Tadmor, and A. Banaszuk. Optimal mixing in recirculation zones. *Physics of Fluids*, pages 867–888, 2004.
[74] S. Ozono. Flow control of vortex shedding by a short splitter plate asymmetrically arranged downstream of a cylinder. *Physics of Fluids*, 11:2928–2934, 1999.
[75] D.S. Park, D.M. Ladd, and E.W. Hendricks. Feedback-control of von karman vortex shedding behind a circular-cylinder at low reynolds-numbers. *Physics of Fluids*, 6:2390–2405, 1994.
[76] A. Pentek, J.B. Kadtke, and G. Pedrizzetti. Dynamical control for capturing vortices near bluff bodies. *Physical Review E*, 58:1883–1898, 1998.

[77] A. Pentek, T. Tel, and Z. Toroczkai. Stabilizing chaotic vortex trajectories: an example of high-dimensional control. *Physics Letters A*, 224:85–92, 1996.

[78] B. Protas. Linear feedback stabilization of laminar vortex shedding based on a point vortex model. *Physics of Fluids*, pages 4473–4488, 2004.

[79] B. Protas and A. Styczek. Optimal rotary control of the cylinder wake in the laminar regime. *Physics of Fluids*, 14:2073–2087, 2002.

[80] D. Rockwell. Vortex-body interactions. *Annual Review of Fluid Mechanics*, 30:199–229, 1998.

[81] H. Sakamoto and H. Haniu. Optimum suppression of fluid forces acting on a circular-cylinder. *Journal of Fluids Engineering-Transactions of The ASME*, 116:221–227, 1994.

[82] G. N. Silva and R. B. Vinter. Necessary conditions for optimal impulsive control problems. *SIAM J. Control Optimization*, 35:1829–1846, 1997.

[83] P.R. Spalart. Airplane trailing vortices. *Annual Review of Fluid Mechanics*, 30:107–138, 1998.

[84] S. Tang and N. Aubry. Suppression of vortex shedding inspired by a low-dimensional model. *Journal of Fluids And Structures*, 14:443–468, 2000.

[85] J.S. Tao, X.Y. Huang XY, and W.K. Chan. Flow visualization study on feedback control of vortex shedding from a circular cylinder. *Journal of Fluids and Structures*, 10:965–970, 1996.

[86] M.S. Triantafyllou, A.H. Techet, Q. Zhu, D.N. Beal, F.S. Hover, and D.K.P. Yue. Vorticity control in fish-like propulsion and maneuvering. *Integrative And Comparative Biology*, 42:1026–1031, 2002.

[87] M.S. Triantafyllou, G.S. Triantafyllou, and D.K.P. Yue. Hydrodynamics of fishlike swimming. *Annu. Rev. Fluid Mech.*, 32:33–53, 2000.

[88] U. G. Vaidya and I. Mezić. Controllability of a class of area-preserving twist maps. UCSB Preprint, 2002.

[89] D.L. Vainchtein and I. Mezić. Control of a vortex pair using a weak external flow. *Journal of Turbulence*, 3, 2002.

[90] D.L. Vainchtein and I. Mezić. Optimal control of a co-rotating vortex pair: Averaging and impulsive control. *Physica D*, 192:63–82, 2004.

[91] Y. Wang, G. Haller, A. Banaszuk, and G. Tadmor. Closed-loop lagrangian separation control in a bluff body shear flow model. *Physics of Fluids*, 15:2251–2266, 2003.

[92] H.M. Warui and N. Fujisawa. Feedback control of vortex shedding from a circular cylinder by cross-flow cylinder oscillations. *Experiments in Fluids*, 21:49–56, 1996.

[93] M.J. Wolfgang, J.M. Anderson, M.A. Grosenbaugh, D.K.P. Yue, and M.S. Triantafyllou. Near-body flow dynamics in swimming fish. *The Journal of Experimental Biology*, 202:2303–2327, 1999.

[94] T. Yang. Impulsive control. *IEEE Transactions on Automatic Control*, 44:1081–1083, 1999.

[95] I. Yasuda and G.R. Flierl. Two-dimensional asymmetric vortex merger: merger dynamics and critical merger distance. *Dynamics of Atmospheres and Oceans*, 26:159–181, 1997.

[96] J.H. Yu and C.F. Driscoll. Diocotron wave echoes in a pure electron plasma. *IEEE Transactions on Plasma Science*, 30:24–25, 2002.

[97] L. Zannetti and A. Iollo. Passive control of the vortex wake past a flat plate at incidence. *Theoretical And Computational Fluid Dynamics*, 16:211–230, 2003.

[98] Q. Zhu, M.J. Wolfgang, D.K.P. Yue, and M.S. Triantafyllou. Three-dimensional flow structures and vorticity control in fish-like swimming. *J. of Fluid Mechanics*, 468:1–28, 2002.

Flow Optimization Using Stochastic Algorithms

Petros Koumoutsakos and Sibylle D. Müller

Institute of Computational Science, ETH Zürich, CH-8092, Switzerland
petros@ethz.ch

Abstract. We present a set of stochastic optimization strategies and we discuss their applications to fluid mechanics problems. The optimization strategies are based on state-of-the-art stochastic algorithms and are extended for the application on fluid dynamics problems. The extensions address the question of parallelization, strategy parameter adaptation, robustness to noise, multiple objective optimization, and the use of empirical models. The applications range from burner design for gas turbines, cylinder drag minimization, aerodynamic profile design, micromixer, microchannel, jet mixing to aircraft trailing vortex destruction.

1 Introduction

The optimization of physical processes for applications in areas such as turbomachinery, aeronautics, and microtechnology poses different challenges to the optimization engineer.

Jet mixing, for example, is an application in which an increased mixing is aimed at in order to reduce noise, suppress signature, and increase lift in civil and/or military aircraft. Mixing the hot jet gas with the surrounding air can be enhanced by actuating the flow at the outlet of the jet. For a systematic search for optimal actuation parameters, one first needs to obtain the mixing rate as a function of the actuation parameters. This can be achieved by setting up experiments or simulations. Let us assume that a computer program is available that simulates the jet and computes the value of the mixing rate. Then, for an automated optimization, we need to wrap a search algorithm around the simulation program. What is a suitable optimization method in this case?

The program can compute the mixing rate as a function of the actuation parameters, that means, only function information is available but no gradient information. To obtain gradients, we would need to (i) approximate them using e.g. finite difference methods that are inaccurate [4], or (ii) compute them using automatic differentiation techniques which is not so trivial and might not be easily applicable to this particular case [3], or (iii) determine them using e.g. an adjoint variable method which is a difficult task that has not been solved yet for jet mixing (and for most other fluid dynamics applications) [36]. The problem of jet mixing exemplifies the difficulties associated with optimization using gradient based methods, namely noise, multiple minima, and the overall absence of an explicit function relating parameters and objective function. The same underlying difficulties are present in other applications tackled in this paper ranging from realistic turbomachinery to shape optimization in nanoscale structures. Therefore, the applicability of gradient methods in these case appears not convincing. More promising for these problems are nongradient optimization techniques despite their slow convergence properties. Among those, we can decide for deterministic or stochastic techniques. As the jet and other fluid dynamics applications represent highly dynamical systems that are susceptible to small changes in actuation parameters, a stochastic optimization algorithm that can handle noise is preferable.

Among stochastic search methods, evolutionary algorithms have become more and more popular in recent years, mainly because of their ease in implementation and their advantages compared with traditional algorithms especially when dealing with nondifferentiable, discontinuous, multimodal and/or noisy optimization functions. As most engineering optimization problems deal with such kinds of functions, it is obvious that evolutionary algorithms are an interesting alternative to classical methods.

Our stochastic optimization framework includes

- the Covariance Matrix Adaptation Evolution Strategy (CMA-ES) [19, 20, 31],
- Evolutionary Algorithms using adaptation techniques that employ Self-Organizing Maps [6, 7, 8, 23],
- Evolutionary Multi-Objective optimization algorithms with adaptation and noise-tolerance [5],
- Clustering Genetic Algorithms for finding correlations in a set of solutions [27], and
- Response Surface Methods including Gaussian processes for interpolation of solutions [1, 16, 24, 41].

The applications are presented in the following order: We start with the experimental optimization of a burner in combustion processes in Sect. 2. Bluff body flow and particularly the minimization of cylinder drag is the topic of Sect. 3. In Sect. 4, we describe how aerodynamic profiles can be designed using novel stochastic optimization concepts. The optimization of micromixers is shown in Sect. 5, microchannel design in Sect. 6, jet mixing in Sect. 7, and aircraft trailing vortex destruction in Sect. 8. Our observations are concluded in Sect. 9.

2 Multi-Objective Optimization in Combustion Processes

The optimization of the combustion process of a stationary gas turbine is a challenging real-world application with conflicting objectives. New governmental laws on emission taxes and global agreements on emission reduction such as the Kyoto resolution on greenhouse-gases (1997, 2001) demand expensive, highly thermodynamically efficient power plants with low emissions. On the other hand, the liberalization of the electric power market puts pressure on overall production costs. In recent years, the use of gas turbines among new power plants has significantly increased due to a number of appealing properties: Using natural gas instead of coal or oil leads to a cleaner combustion, while moderate installation and operating costs and a high thermodynamically efficiency reduce overall energy production costs. Moreover, using the exhaust heat for a steam turbine in a combined cycle is one way to increase power output and efficiency of the plant.

A central component in the design of a gas turbine is the design of the burners in the combustion chamber. The burners mix air and fuel and combust them continuously. This is different to Diesel engines, which combust in a cyclic manner. The design of a burner addresses two main objectives: First, the burner should mix air and fuel uniformly for low emissions, since the presence of areas of rich combustion results in increased NO_x emissions and a non-homogeneous temperature distribution may damage the turbine blades. Second, the burner should produce a stable combustion flame, avoiding undesired pulsations. Pulsations are due to thermo acoustic waves, which occur in particular for lean combustion when operating under part load condition. They reduce the lifetime of the turbine by fatigue and by destroying the film cooling along the blades surface. These two objectives are conflicting, thus motivating the requirement for a variety of designs as manifested on a Pareto front. The lack of viable analytical models and the limited information about the underlying physical processes involved makes combustion processes a suitable candidate for the optimization using stochastic optimization techniques [13].

Our contribution has been the application of multi-objective evolutionary optimization to a realistic industrial set-up [5]. Experimental setups present a number of challenges to any optimization technique including: availability only of pointwise information, experimental noise in the objective function, uncontrolled changing of environmental conditions and measurement failure. Based upen the SPEA [44], our evolutionary algorithm incorporates a number of new concepts, as dictated by the experiments, such as domination dependent lifetime, re-evaluation of solutions, and modifications in the update of the archive population.

We consider the optimization of a single burner in an atmospheric test-rig. Preheated air enters the test-rig from the plenum chamber and is mixed with fuel in the low-emission burner by swirl. The burner stabilizes the combustion flame in a predefined combustion area by a controlled vortex breakdown. The fuel is natural gas or oil and is injected through injection holes, which are uniformly distributed along the burner. A detailed description is given by Jansohn et al. [21]. Various investigations aimed to reduce pulsations and emissions of the burner by active and passive control mechanisms. We consider a passive control mechanism, choosing the fuel flow rates through the injection holes of the burner as design variables of the setup, due to the low modification cost for the gas turbine compared to an active control system. Eight continuous valves are used to control the fuel rates. Each valve controls the mass flow through a set of adjacent injection holes along the burner axis.

The Pareto front is constructed for the objectives of minimization of NO_x emissions and reduction of the pressure fluctuations of the flame, yielding reduced emissions and pulsation of the burner [5]. The results from this work have led to three patents for new burner designs [13, 14].

3 Cylinder Drag Minimization

A real coded genetic algorithm was implemented for the optimization of actuator parameters for cylinder drag minimization. We consider the two-dimensional and incompressible flow at $Re = 500$ past a circular cylinder, in combination with two types of idealized actuators that are allowed either to move steadily and tangentially to the cylinder surface ("belts"), or to steadily blow/suck with a zero net mass constraint. The genetic algorithm that we implemented has the property of identifying minima basins, rather than single optimum points. The knowledge of the shape of the minimum basin enables further insights in the system properties and provides a sensitivity analysis in a fully automated way. The drag minimization problem is formulated as an optimal regulation problem.

By means of the clustering property of the present genetic algorithm, a set of solutions producing drag reduction of up to 50% is identified. A thorough cluster analysis [27] revealed that the important parameters for the flow control are only the ones corresponding to actuators containing the separation point in the uncontrolled flow. At the same time all the other actuators could be sliding/blowing/sucking with random velocities or remain fixed. To verify this hypothesis another validation run was performed, this time maintaining active only the relevant actuators.

A comparison between the two types of actuators, based on the clustering property of the algorithm indicates that blowing/suction actuation parameters are associated with larger tolerances when compared to optimal parameters for the belt actuators. The possibility to use few strategically placed actuators in order to obtain a significant drag reduction was explored using the clustering diagnostics of this method. The optimal belt-actuator parameters obtained by optimizing the two-dimensional case have been employed in three dimensional simulations, by extending the actuators across the span of the cylinder surface. The three dimensional controlled flow exhibits a strong two-dimensional character near the cylinder surface, resulting in significant drag reduction [27].

The results obtained using two dimensional simulations are shown to be useful for three dimensions when the actuators are suitably extended on the third dimension of the flow [27]. This suggests that optimization in two dimensions followed by a validation of the results in three dimensions is a viable approach to the rapid design of realistic control devices.

4 Aerodynamic Profile Design

We consider the automated profile design for compressor blades of stationary gas turbines. The design is restricted to blades with subsonic flow. An optimization loop is implemented comprising an optimization algorithm, a profile generation tool and a computational fluid dynamics (CFD) analysis tool. The profile generator describes the profile by a set of Bezier splines whose control points are encoded into engineering design parameter like the profile length, the nose and trailing edge radius, and the curvature distribution [22, 42]. The flow analysis is performed with MISES [15], a quasi 3D computational fluid dynamics solver, which solves the Euler equation with an integral, viscous boundary layer formulation. It takes into account the change in the streamline thickness along the profile (quasi 3D). Our approach is to calculate various incidences in order to approximate the loss polar of the profile as given in Fig. 2. The loss polar specifies the behavior of the profile over the complete operating range. A disadvantage is the large number of flow calculations, which are needed to specify the polar as in the optimization of [22]. Furthermore, there is the problem of how many incidences should be computed and for which values.

In the following, we do not compute the complete loss polar and show that it is sufficient to compute 3 different incidences in order to assess a profile. The 3 calculations are performed for the design condition, i.e. $0°$ incidence and for one positive incidence $I_1 > 0°$ and one negative incidence $I_2 < 0°$. The key concept is to define I_1 and I_2 by a free multiplier θ as $I_1 = 1.0 \cdot \theta$ and $I_2 = -0.8 \cdot \theta$. This definition takes into account that the positive incidence I_1 is more critical for stall than I_2. The incidence multiplier θ is an additional design variable. The profile losses for the 3 incidences are summed to the first objective function f_1. For small values of θ, the losses are computed at small incidences. An optimization for small values of θ leads profiles which have minimal losses in the vicinity of the design condition, while for large values of θ, profiles are optimized for a large incidence range. Thus, θ is not only used as free design variable, but also as second objective function f_2. We minimize f_1 and maximize f_2 where the objective functions include penalties: $f_1 = \sum_{i=1}^{3} l_i + p_1 + p_2 + p_3 + p_4$ and $f_2 = \theta - p_1$, where l_i is the profile loss for the incidence i and p_1 to p_4 are 4 penalties, which are non-zero, if the corresponding constraint is violated. The first p_1 penalty regards convergence of the CFD solver. Penalty p_2 to p_4 address flow separation and mechanical stresses. The 15 free design variables are the parameters from the profile generator and the incidence multiplier θ. Two optimization runs are performed for a profile design at an inlet Mach number of 0.67, a desired flow turning of 12° and $\delta\beta = 0.1°$. In the first optimization, the two conflicting objectives f_1 and and f_2 are aggregated and a single objective algorithm is used. The second optimization run is a Pareto optimization for the two conflicting objectives. Most optimization algorithm are designed for a single objective function. Thus, for considering multiple objectives, the objectives have to be aggregated into a single figure of merit f_M, which is then optimized. Here, we restrict ourselves to minimization of the figure of merit and construct it as $f_M = f_1 - f_2$.

We compare the convergence properties of the CMA evolution strategy and the optimization algorithm including a Gaussian Process model, respectively. A separete Gaussian Process is constructed for the loss at each design incidence as well as for each constraint and the prediction of all models is aggregation in order to approximate the merit function. First, 100 solutions are computed randomly and then the model is used to search for promising solutions. The model is always trained with all currently evaluated solutions.

In Fig. 1, the merit function is plotted over the number of design evaluations N. The CMA-ES converges by constantly decreasing the merit function and $f_M = -1.6868$ is obtained as best function value after 1000 function evaluations. In the figure, the merit function for the algorithm using the model decreases by a large value as the model is firstly used at $N = 100$ evaluations. The initial 100 random solutions are already sufficient to approximate the merit function well. After $N = 300$ evaluations, the best function value

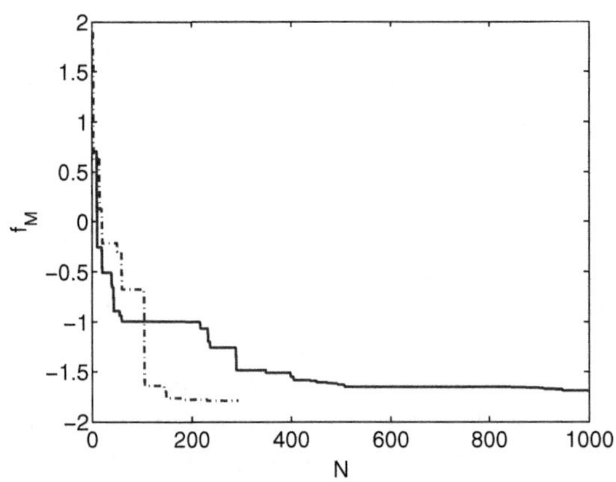

Fig. 1. The merit function versus the number of design evaluations for the CMA (–) and for the Gaussian process model (- - -)

is $f_M = -1.7892$. The plot shows the superior performance of the Gaussian process model compared with the CMA-ES.

We consider the optimization of the two objectives as a Pareto optimization problem. The first objective f_1 contains the losses, which is to be minimized. The second objective f_2 is the incidence multiplier and is to be maximized. The Pareto front underlines the conflict in optimizing the two objectives. For small incidence multipliers, the losses are low, since all 3 incidences are computed almost at the design point. For large incidence multipliers, the loss increases for two reasons. First, the flow is computed at larger incidences leading to higher losses and second, the profile losses are higher at the design condition, since the design has to be more robust for converging at the high incidences.

A multi-objective evolutionary algorithm with adaptive recombination and mutation operators is used for the Pareto optimization [7]. In total, 10.000 solutions are evaluated. Among all evaluated solutions, 5.461 solutions do not violate any constraints and generate a Pareto front of 283 solutions (Fig. 2). Two Pareto solutions are marked in the figure and their loss polar is given in Fig. 2. The minimal losses are at about 1.4 %. The attainable operating range is considered to be bounded by the double of the minimal losses [22]. Solution A contains the smaller incidence multiplier and the loss polar shows lower losses close to the design incidence than solution B, but comprises a smaller operating range. For solutions A and B, the operating range is about 14.4° and 15.5°, respectively. Both polars are characterized by a smooth and continuous increase of losses over the absolute incidence. This indicates a

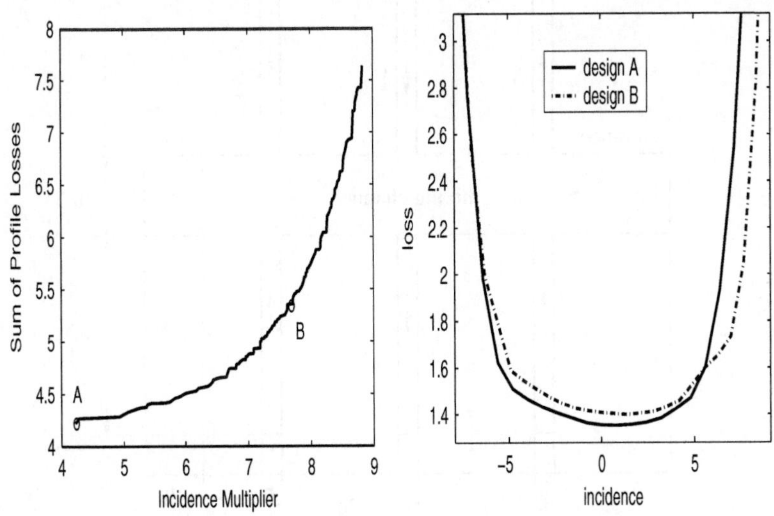

Fig. 2. Pareto front [*left*] for the profile optimization, and loss polar [*right*] for two selected Pareto solutions

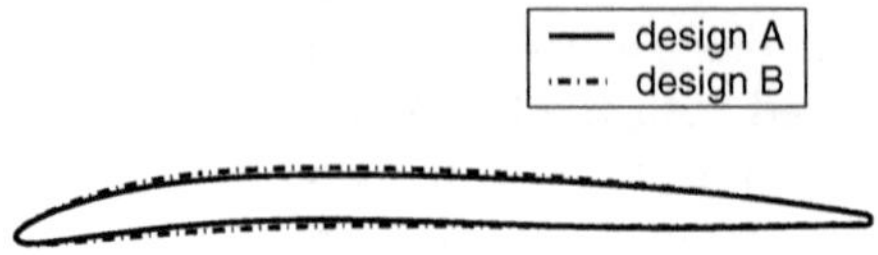

Fig. 3. Profile shape for the two selected Pareto solutions

soft stall behavior. Figure 3 contains the profile shape. Solution A shows the smaller nose radius as well as the smaller maximal thickness.

5 Micromixer

We studied mixing in a transverse-momentum micromixer for pharmaceutical applications. The mixer involves the parallel injection of two fluids which in an uncontrolled configuration do not mix due to the low Reynolds number of the flow. The flow configuration is shown in Fig. 4.

The control involves the use of side micropistons which should be activated so as to induce mixing. A straightforward trial and error experimentation with these actuations did not lead to any significant mixing. Extensive theoretical studies have identified suitable actuation parameters.

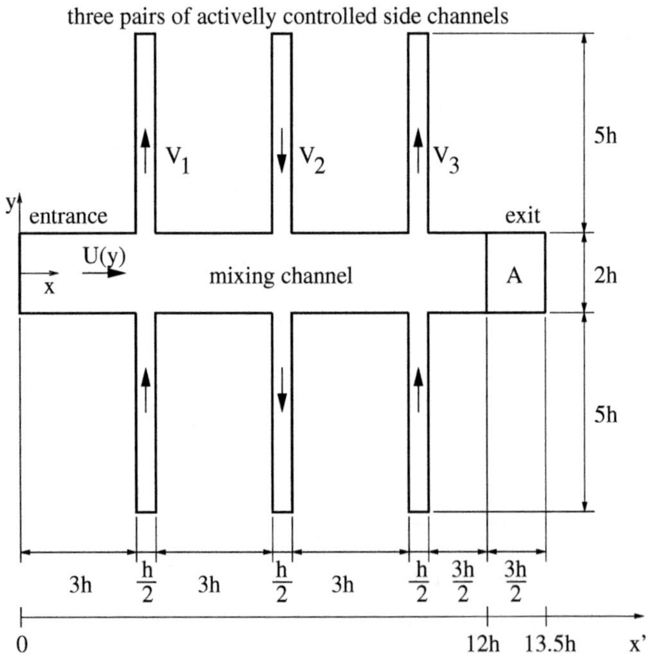

Fig. 4. Sketch of the flow configuration

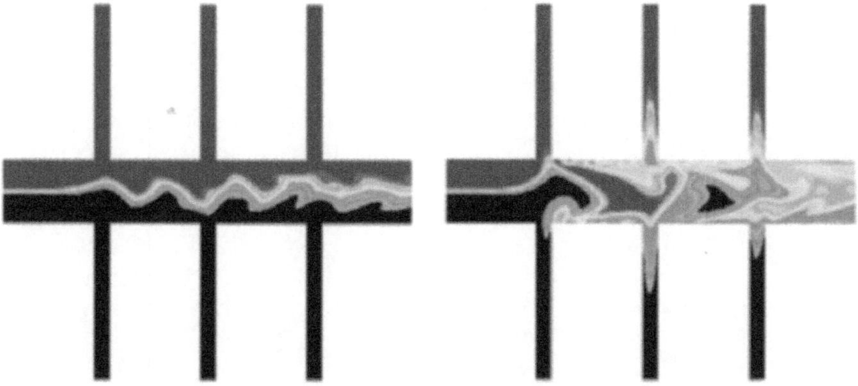

Fig. 5. Flow actuated by the initial frequencies $\boldsymbol{x} = (1/2, 1/2, 1/2)$ (*left*) and by the optimal frequencies $\boldsymbol{x} = (0.14, 0.32, 0.50)$ (*right*)

Our approach was to combine evolution strategies as optimization method with the simulation of the mixing behavior of the two fluids. The flow is modeled by the Navier-Stokes and convection-diffusion equations discretized using a second order finite volume technique and solved on a Cartesian grid using a standard computational fluid dynamics package [40]. The chosen optimization strategy was an evolution strategy with covariance matrix adaptation [17, 18, 19]. Optimization parameters are the frequencies of the movement of the micropistons and the objective is to increase the mixing of the two fluids which is estimated from the local variance of the concentration field.

Figure 5 shows two snapshots of the flow in the micromixer at time $t = 45$ for initial and optimal frequencies, respectively.

It was shown that the evolution algorithms can identify, in an automated fashion, effective actuations with mixing results that far exceeded those obtained by theoretical studies for the same configuration. In addition, we found that optimal frequencies for an increasing number of transverse channels are superposable despite the nonlinear nature of the mixing process [29, 30, 32].

6 Microchannel Flow

We apply both single and multiobjective EAs applied to a fluidic microchannel design problem [37, 38]. Bio-analytical applications require long thin channels for DNA sequencing by means of electrophoresis. In order to pack a channel of several meters in length onto a small square plate, curved geometries are required. However, curved channels introduce dispersion and, therefore, limit the separation efficiency of the system. The question is how to shape the contour of the channel in order to minimize dispersion. A detailed description of the problem as well as an optimization solution using gradient methods can be found in [28].

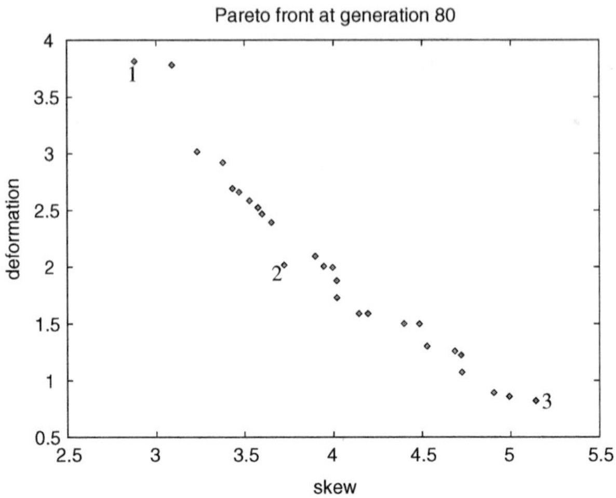

Fig. 6. Pareto-front of nondominated solutions after 80 generations

In a first study, we study a single-objective: The goal is to minimize the final skew of the flow inside the channel, i.e., it is required that the iso-values of the advected species be normal to the flow field when they exit the channel. The shape of the 90-degree turn is described by 11 parameters. The main result of our optimization using a $(1 + 1)$-ES with $1/5$ success rule is that we find a novel double-bump shape that has not been found before. Previous studies had shown single-dented designs always.

In a second study, we optimize two goals, the skew and the total deformation of the channel contour. The second goal is introduced to take into account manufacturing costs which increase when the channel is deformed strongly. The results from our two-objective minimization using the Strength Pareto Evolutionary Algorithm [44] are compared with the gradient-based optimization results by [28]. Figure 6 shows the Pareto-optimal trade-off front after 80 generations of the algorithm, and Fig. 7 shows the corresponding solutions, i.e., optimized shapes of the channel. From this front, we can decide for a solution with minimal skew at the expense of a higher deformation (point 1; represented in Fig. 7, left), some intermediate result (point 2), or with minimal deformation with the lowest skew possible (point 3, represented in Fig. 7, right).

Figure 8 shows two classes of optimized shapes obtained by [28] using gradient methods. Interestingly, the gradient technique offers two different designs, namely the single-dented (Fig. 8, left) and the double-dented shapes (Fig. 8, right) which we found with the evolution strategy also. Therefore, we obtain qualitatively similar results from both methods. Using the gradient method, the skew is reduced by one order of magnitude [28] which is comparable to the numbers obtained by evolutionary optimization. While trial and error procedures were used in the gradient methods to obtain various

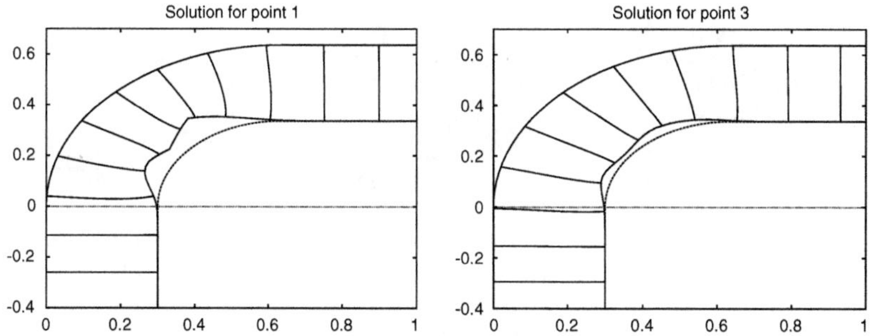

Fig. 7. Solution at point 1 (*left*) and at point 3 (*right*)

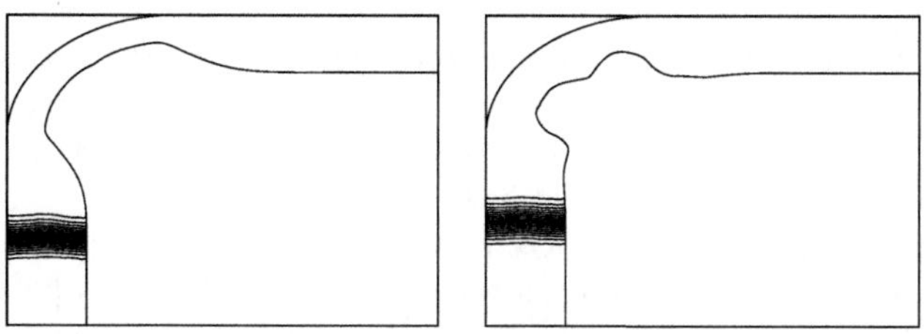

Fig. 8. First (left) and second (right) optimized shape using a gradient method [28]

solutions, ES provides us with a number of solutions and a Pareto front in an automated fashion. Unlike the gradient based methods which require an explicit formulation of the optimization problem at hand, the evolution strategy provides a straightforward optimization procedure.

7 Jet Mixing

Enhanced jet mixing has several technological applications with the goal of improving safety, efficiency, or reliability. One example relevant to military aircraft is signature suppression where the dispersion of the hot jet exhaust is aimed at. Another example in jet propulsion is to reduce the plume temperature on aerodynamic surfaces, such as the blown flap of a C-17 aircraft. In combustion processes, it is often important to enhance the turbulent mixing of the chemical components to make the combustion process more efficient with size and weight reductions possible, and to reduce the concentration of pollutants. The mixing rate of a jet can be significantly altered by applying a suitable excitation at the jet orifice. Since the external forcing interacts with the natural modes of the jet in a nonlinear way, it is difficult to estimate which kind of actuation is optimal to increase mixing.

Our work has focused on jet optimization using DNS and LES of an incompressible jet [30, 32]. We have studied helical and combined helical and axial forcing of a jet that maximizes mixing by combining an evolution strategy

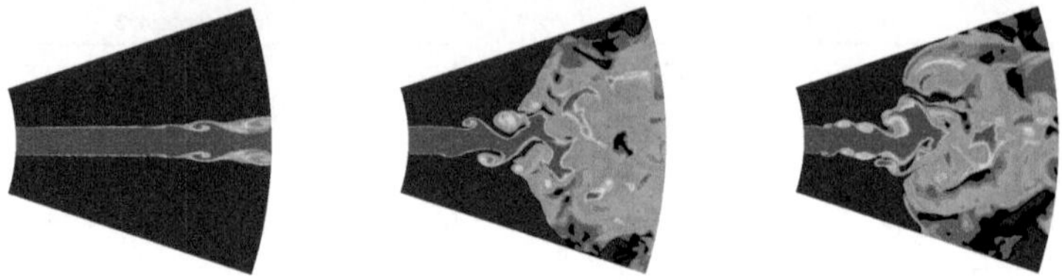

Fig. 9. Jet at Re = 1500. Left: No actuation. Middle: With maximum radial velocity for one-frequency excitation; Strouhal frequency $\mathrm{St}_h = 0.30$, helical amplitude $A_h = 0.08$. Right: With maximum radial velocity for two-frequency excitation; axial Strouhal frequency $\mathrm{St}_a = 0.72$, helical Strouhal frequency $\mathrm{St}_h = 0.29$, axial amplitude $A_a = 0.025$, helical amplitude $A_h = 0.075$

with direct numerical simulation of a round jet. Varying the actuation at the orifice, we searched for the forcing that maximizes various metrics related to enhanced jet mixing. For one- and two-frequency actuation, we have obtained optimum mixing results presented in Fig. 9.

Since DNS of high Re flows were becoming exceedingly expensive, we considered vortex filament methods as a model for high Re flows [9]. The three-dimensional evolution of a nominally axisymmetric jet subject to azimuthal and helical perturbation waves has been studied [25]. Circular filaments are uniformly initialized, with uniform circulation, and then slightly displaced to generate the perturbation. We performed evolutionary optimization, starting from values given by [26], that resulted in an improvement of the total length of the filaments (the mixing metric) of 42 % compared to the initial values.

The main finding from the jet mixing optimization using both vortex models and direct simulations was that the upper bound of the actuation amplitude appears to be of utmost relevance to mixing. A study of the physics of mixing has yet to reveal why this behavior emerges.

We have extended the implementation of an evolution strategy to the experimental laboratory at Stanford University [37]. We have used the original device developed by Parekh and Reynolds to study bifurcating jets. By automating the control process using LABVIEW™ and by coupling this process to our optimization algorithm we have obtained a 50% enhancement in the temperature mixing profile as compared to the classic experiments [33].

8 Aircraft Trailing Vortex Destruction

Trailing vortices are naturally shed by airplanes. They result in a strong downwash which extends for several miles behind the plane and poses a hazard to following aircraft, in particular at take-off and landing. Several previous studies propose to alleviate the hazard by introducing perturbations to trigger instabilities, and ultimately, break up the vortices [2, 12].

Recent studies [10, 35] have considered instabilities unique to several pairs of vortices which model aircraft wakes in landing configuration [39]. Some of

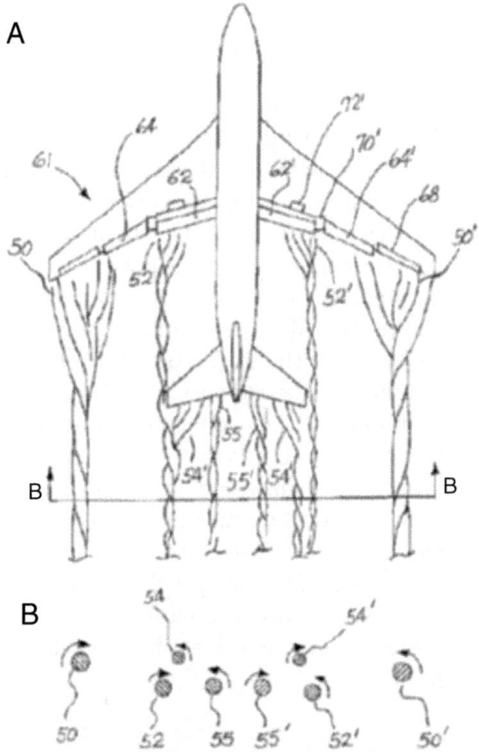

Fig. 10. Sketch of vortex system shed by an airplane (Courtesy of Crouch and Spalart [11]). B is a cross section of A as shown

these vortices quickly merge, but others persist for long times. At a distance of several spans behind a typical airplane, three persistent vortex pairs can generally be observed, originating at the tips of the wings, the outboard flaps, and the fuselage, see Fig. 10.

Crouch [10] has studied the linear stability of two pairs of co-rotating vortices (tip and outboard flap). He identified several instability modes depending on the angle, wavelength, and amplitudes of the perturbations that are imparted to each pair. Although the points of view adopted in recent studies [10, 11, 35] differ in several respects, in particular in the way the instability growth is measured, they have in common that they can provide us with a better understanding of the mechanisms by which the cooperative instabilities of several pairs can result in enhanced growth rates. Moreover, the configurations studied in these works are investigated with a view to implementing them in actual wing designs. One of the findings reported in [10] and [35] is the extreme sensitivity of the overall dynamics with respect to the initial state of the vortex pairs. In [10], the most effective transient growth was achieved when the outboard pair was not initially perturbed, while in [35] early reconnection was obtained for a particular value of the inboard vortices separation.

This motivates our attempt to perform a more systematic parameter search and identify the wake system which would produce the largest instability growth. Our goal is to revisit the above studies using viscous vortex methods and optimization with evolution strategies. Vortex methods are well adapted to wake simulation as they require the discretization of only the region of vorticity [9]. Note that the work of [35] is in part based on a vortex filament method. Using a $(1+1)$-evolution strategy, we optimize a total of seven parameters describing the perturbation of two pairs of co-rotating vortices, the tip and outboard vortices, and the geometry. The objective function to be maximized is the instability on the tip vortex.

We compare the results from the evolutionary optimization with parameters reported in [10] as leading to efficient transient growth. Some striking similarities can be noticed between these two sets of parameters. In particular, the ES has selected perturbations that are mostly located on the tip vortex, confirming the observation from linear stability analysis in [10] of efficient transient growth when the outboard flap vortex is unperturbed. The wavelengths of the perturbations are also close to the ones given in [10]. The case of four pairs of vortices is also considered and leads to a larger distortion of the tip vortex [37].

9 Summary and Conclusions

Biologically inspired stochastic search algorithms were applied to a variety of engineering problems ranging from aerodynamics and turbomachinery to microtechnology. In summary, the results show that these optimization methods are highly suitable for optimization in applications that are characterized by noise, multimodality, and no availability of gradient information.

References

[1] El-Beltagy, M.A., Keane, A.J., "Evolutionary optimization for computationally expensive problems using gaussian processes," Hamid Arabnia, editor, *Proc. Int. Conf. on Artificial Intelligence IC-AI'2001, Las Vegas*, pp. 708–714. CSREA Press, 2001.

[2] Bilanin, A.J., Widnall, S.E., "Aircraft wake dissipation by sinusoidal instability and vortex breakdown," AIAA 73–107, 1973.

[3] Bischof, C., Griewank, A., "Tools for the automatic differentiation of computer programs," *ICIAM/GAMM 95: Issue 1. Numerical Analysis, Scientific Computing, Computer Science*, pp. 267–272, 1996.

[4] Booker, A.J.J., Dennis Jr, J., Frank, P.D., Serafini, D., Torczon, V., Trosset, M.W., "A rigorous framework for optimization of expensive functions by surrogates," *ICASE report No. 98-47*, NASA Langley Research Center, Hampton, VA, 1998.

[5] Büche, D., Stoll, P., Dornberger, R., Koumoutsakos, P. "Multi-objective Evolutionary Algorithm for the Optimization of Noisy Combustion Problems," *IEEE Transactions on Systems, Man, and Cybernetics, Part C*, Vol. 32, No. 4, pp. 460–473, 2003.

[6] Büche, D., Guidati, G., Stoll, P., Koumoutsakos, P. "Self-organizing maps for pareto optimization of airfoils," *Seventh International Conference on Parallel Problem Solving from Nature (PPSN VII)*, Granada, Spain. Springer-Verlag, 2002.

[7] Büche, D., Milano, M., Koumoutsakos, P., "Self-Organizing Maps for Multi-Objective Optimization," *Workshop Proceedings of the Genetic and Evolutionary Computation Conference (GECCO)*, Morgan Kaufmann Publishers, San Francisco, CA, pp. 152–155, 2002.

[8] Büche, D., Milano, M., Koumoutsakos, P., "Growing Self-Organizing Maps for Multi-Objective Optimization," Evolutionary Computation Journal, 2002 (submitted).

[9] Cottet, G.H., Koumoutsakos P.D., "Vortex Methods – Theory and Practice," Cambridge University Press, 2000.

[10] Crouch, J.D., "Instability and transient growth for two trailing-vortex pairs," *Journal of Fluid Mechanics*, No. 350, pp. 311–330, 1997.

[11] Crouch, J.D., Spalart, P.R., "Active system for early destruction of trailing vortices," *US Patent 6082679*, issued July 4, 2000.

[12] Crow, S.C., Bate, E.R., "Lifespan of trailing vortices in a turbulent atmosphere," *Journal of Aircraft*, No. 13, pp. 476–482, 1997.

[13] Dornberger, R., Stoll, P., Paschereit, Ch.O., Schuermans, B., Bueche, D., Koumoutsakos, P., "Numerisch experimentelle Optimierung des Mischungsprofils für die Kontrolle verbrennungsgetriebener Schwingungen und Emissionen", *German Patent No. 101 04 151.9*, ALSTOM Power (Schweiz) AG, 2001.

[14] Dornberger, R., Stoll, P., Paschereit, Ch.O., Schuermans, B., Bueche, D., Koumoutsakos, P., "Redundanzfreier Ansteuerungsmechanismus der Ventile zur Beeinflussung des Mischungsprofiles für die Kontrolle verbrennungsgetriebener Schwingungen, *German Patent No. 101 04 150.0*, ALSTOM Power (Schweiz) AG, 2001.

[15] Drela, M., Youngren, H., "A User's Guide to MISES 2.53," MIT, 1998.

[16] Gibbs, M.N., MacKay, D.J.C., "Manual for tpros, v2.0," 1997.

[17] Hansen, N., Ostermeier, A., "Convergence Properties of Evolution Strategies with the Derandomized Covariance Matrix Adaptation: The $(\mu/\mu_I, \lambda)$-CMA-ES," *Proceedings of the 5th European Congress on Intelligent Techniques and Soft Computing (EUFIT'97)*, pp. 650–654, 1997.

[18] Hansen, N., "Verallgemeinerte individuelle Schrittweitenregelung in der Evolutionsstrategie: Eine Untersuchung zur entstochastisierten, koordinatensystemunabhängigen Adaptation der Mutationsverteilung," Mensch & Buch Verlag, Berlin, 1998.

[19] Hansen, N., Ostermeier, A., "Completely Derandomized Self-Adaptation in Evolution Strategies," *Evolutionary Computation*, Vol. 9, No. 2, pp. 159–195, 2001.

[20] Hansen, N., Müller, S.D., Koumoutsakos, P., "Reducing the Time Complexity of the Derandomized Evolution Strategy with Covariance Matrix Adaptation (CMA-ES)," *Evolutionary Computation Journal*, MIT Press, Vol. 11, No. 1, 2003.

[21] Jansohn, P., Ruck, T., Steinbach, C., Knöpfel, H.-P., T. Sattelmayer, T., "Development of the Advanced EV (AEV) Burner for the ABB GTX 100 Gas Turbine", *ASME Turbo Asia 97*, Singapore, 1997.

[22] Köller, U., Mönig, R., Küsters, B., Schreiber, H.-A., "Development of Advanced Compressor Airfoils for Heavy-Duty Gas Turbines, Part I: Design and Optimization," *ASME Journal of Turbomachinery*, Vol. 122, No. 3, pp. 397–405, 1999.

[23] Kohonen, T., "Self-organizing maps." *Springer series in information sciences*, 3rd edition, 2001.

[24] "Gaussian processes - a replacement for supervised neural networks," *Lecture notes for a tutorial at NIPS*, 1997.

[25] Martin, J.E., Meiburg, E., "Numerical investigation of three-dimensionally evolving jets under helical perturbations," *Journal of Fluid Mechanics,* Vol. 243, pp. 457–487, 1992.

[26] Martin, J.E., *Personal communication,* 2001.

[27] Milano, M., Koumoutsakos, P., "A Clustering Genetic Algorithm for Cylinder Drag Optimization," *Journal of Computational Physics*, Vol. 175, pp. 79–107, 2002.

[28] Mohammadi, B., Molho, J.I., Santiago, J.A., *Design of Minimal Dispersion Fluidic Channels in a CAD-Free Framework*, Center for Turbulence Research Annual Research Briefs, 2000.

[29] Müller, S.D., Sbalzarini, I.F., Walther, J.H., Koumoutsakos, P., "Evolution Strategies for the Optimization of Microdevices," *Proceedings of the Congress on Evolutionary Computation (CEC 2001)*, pp. 302–309, 2001.

[30] Müller, S.D., Koumoutsakos, P., "Mixing Optimization with Evolution Strategy," *Evolutionary Methods for Design, Optimization and Control with Applications to Industrial Problems – Proceedings of the EUROGEN 2001 Conference*, pp. 448-452, September 2001.

[31] Müller, S.D., Hansen, N., Koumoutsakos, P., "Improving the Serial and Parallel Performance of the CMA-Evolution Strategy with Large Populations," *Proceedings of the Seventh International Conference on Parallel Problem Solving from Nature (PPSN VII)*, Lecture Notes in Computer Science, Vol. 2439, Springer, Berlin, pp. 422–431, 2002.

[32] Müller, S.D., Koumoutsakos, P., "Control of Micromixers, Jets, and Turbine Cooling Using Evolution Strategies," *Manipulation and Control of Transverse Jets*, International Centre for Mechanical Sciences (CISM), Udine, Italy, 2002.

[33] Parekh, D., Leonard, A., Reynolds, W.C., "Bifurcating of Round air jets by dual-mode acoustic excitation," AIAA 87–0164, 1987.

[34] Rennich, S.C., "Accelerated destruction of aircraft wake vortices," *Ph.D. thesis*, Stanford University, 1997.

[35] Rennich, S.C., Lele, S.K., "A method for accelerating the destruction of aircraft wake vortices." AIAA 98–0667, 1998.

[36] Reuther, J.J., Alonso, J.J., Jameson, A., Rimlinger, M.J., Saunders, D., "Constrained multipoint aerodynamic shape optimization using an adjoint

formulation and parallel computers, Part I and II," AIAA Journal of Aircraft, Vol. 36, No. 1, pp. 51–71, 1999.

[37] Sbalzarini, I.F., Müller, S.D., Koumoutsakos, P., Cottet, G.-H., "Evolution Strategies for Computational and Experimental Fluid Dynamics Applications," *Proceedings of the Genetic and Evolutionary Computation Conference (GECCO)*, Spector, L., Goodman, A., Wu, A., Langdon, W.B., Voigt, H.-M., Gen, M., Sen, S., Dorigo, M., Pezeshk, S., Garzon, M., Burke, E. (eds.), Morgan Kaufmann Publishers, San Francisco, CA, pp. 1064–1071, 2001.

[38] Sbalzarini, I.F., Müller, S.D., Koumoutsakos, "Optimization Using Multiobjective Evolution Strategies," *Proceedings of the First International Conference on Evolutionary Multi-Objective Optimization (EMO'01)*, Lecture Notes in Computer Science, Vol. 1993, Springer, Berlin, 2001.

[39] Spalart, P.R., "Airplane trailing vortices," Annual Review of Fluid Mechanics, No. 30, pp. 107–138, 1998.

[40] *Star-CD User Manual,* Computational Dynamics Ltd., London, 1997.

[41] Torczon, V., Trosset, M.W., "Using approximations to accelerate engineering design optimization," *ICASE Report No. 98–33*, Technical report, NASA Langley Research Center Hampton, VA, 23681-2199, 1998.

[42] Trigg, M.A., Tubby, G.R., Sheard, A.G., "Automatic Genetic Optimization Approach to Two-Dimensional Blade Profile Design for Steam Turbines," *ASME Journal of Turbomachinery*, Vol. 121, No. 1, pp. 11–17, 1999.

[43] Volpert, M., Mezic, I., Meinhart, C.D., Dahleh, M., "Modeling and Analysis of Mixing in an Actively Controlled Micromixer," *Unpublished report, University of Santa Barbara, CA*, 2000.

[44] Zitzler, E., Thiele, L., "Multiobjective Evolutionary Algorithms: A Comparative Case Study and the Strength Pareto Approach," *IEEE Transactions on Evolutionary Computation*, Vol. 3, No. 4, pp. 257–271, 1999.

GPSR Compliance

The European Union's (EU) General Product Safety Regulation (GPSR) is a set of rules that requires consumer products to be safe and our obligations to ensure this.

If you have any concerns about our products, you can contact us on ProductSafety@springernature.com

In case Publisher is established outside the EU, the EU authorized representative is:

Springer Nature Customer Service Center GmbH
Europaplatz 3
69115 Heidelberg, Germany

Batch number: 10704695

Printed by Printforce, the Netherlands